Introduction to Engineering Mathematics

Introduction to Engineering Mathematics

Adrian Day

WILLFORD **P**RESS

www.willfordpress.com

Published by Willford Press,
118-35 Queens Blvd., Suite 400,
Forest Hills, NY 11375, USA

ISBN: 978-1-64728-332-2

Cataloging-in-Publication Data

Introduction to engineering mathematics / Adrian Day.
    p. cm.
Includes bibliographical references and index.
ISBN 978-1-64728-332-2
1. Engineering mathematics. 2. Mathematical analysis. I. Day, Adrian.
TA330 .I58 2022
620.001 51--dc23

For information on all Willford Press publications
visit our website at www.willfordpress.com

WILLFORD PRESS

TABLE OF CONTENTS

Permissions

Index

It is with great pleasure that I present this book. It has been carefully written after numerous discussions with my peers and other practitioners of the field. I would like to take this opportunity to thank my family and friends who have been extremely supporting at every step in my life.

The branch of applied mathematics that is concerned with the utilization of mathematical methods and techniques in engineering and industry is referred to as engineering mathematics. It is an interdisciplinary subject which is closely related to other fields such as engineering physics and engineering geology. Some of the major areas of study within this field are differential equations, real and complex analysis, approximation theory, Fourier analysis and potential theory. There are various specializations within this field such as engineering optimization and engineering statistics. Engineering statistics involves the study of data related to numerous manufacturing processes like tolerances, type material and fabrication process control. Engineering optimization uses optimization techniques for achieving the design goals in engineering. The topics included in this book on engineering mathematics are of utmost significance and bound to provide incredible insights to readers. It is a compilation of chapters that discuss the most vital concepts in this field. This book is an essential guide for both academicians and those who wish to pursue this discipline further.

The chapters below are organized to facilitate a comprehensive understanding of the subject:

Chapter – Introduction

The branch of applied mathematics which is primarily concerned with the application of mathematical methods and techniques in engineering and industry is referred to as engineering mathematics. It includes topics such as vector algebra and complex numbers. This is an introductory chapter which will introduce briefly all the significant aspects of engineering mathematics.

Chapter – Differential and Integral Calculus

The subfield of calculus which deals with the study of the rates at which quantities change is known as differential calculus. Integral calculus studies the notion and application of integrals and focuses on total size or value, such as lengths, areas, and volumes. The topics elaborated in this chapter will help in gaining a better perspective about these branches of mathematics.

Chapter – Vector Calculus

Vector calculus, also known as vector analysis, is a mathematical branch that focuses on the differentiation and integration of vector fields. The main concepts used in vector calculus are Strokes' theorem, divergence theorem, curl, etc. The chapter closely examines the concepts of vector calculus to provide an extensive understanding of the subject.

Chapter – Differential Equations

Differential equations are the mathematical equations that relate some functions with their derivatives. They are divided into ordinary differential equations and partial differential equations. The topics elaborated in this chapter will help in gaining a better perspective about these types of differential equation.

Chapter – Matrix

A rectangular array of numbers, symbols and expressions arranged in rows and columns is known as a matrix. Some of the focus areas of matrix are determinant, invertible matrix, Cayley-Hamilton theorem, LU decompositions, Eigen values and vectors, etc. These diverse areas of matrix have been thoroughly discussed in this chapter.

Chapter – Complex Analysis

The branch of mathematical analysis that examines functions of complex numbers is referred to as complex analysis. Some of the main concepts of complex analysis are Cauchy's integral theorem and residue theorem, complex function, analytic function, etc. This chapter has been carefully written to provide an easy understanding of various aspects of complex analysis.

Adrian Day

Introduction

The branch of applied mathematics which is primarily concerned with the application of mathematical methods and techniques in engineering and industry is referred to as engineering mathematics. It includes topics such as vector algebra and complex numbers. This is an introductory chapter which will introduce briefly all the significant aspects of engineering mathematics.

Mathematics is the art of applying mathematics to complex real-world problems. It combines mathematical theory, practical engineering and scientific computing to address today's technological challenges.

Engineering Mathematics is a creative and exciting discipline, spanning traditional boundaries. Engineering mathematicians can be found in an extraordinarily wide range of careers, from designing next generation Formula One cars to working at the cutting edge of robotics, from running their own business creating new autonomous vehicles to developing innovative indices for leading global financial institutions.

VECTOR ALGEBRA

Scalars

A physical quantity which is completely described by a single real number is called a scalar. Physically, it is something which has a magnitude, and is completely described by this magnitude. Examples are temperature, density and mass.

Vectors

The concept of the vector is used to describe physical quantities which have both a magnitude and a direction associated with them. Examples are force, velocity, displacement and acceleration.

Geometrically, a vector is represented by an arrow; the arrow defines the direction of the vector and the magnitude of the vector is represented by the length of the arrow, figure.

Analytically, vectors will be represented by lowercase bold-face Latin letters, e.g. a, r, q.

The magnitude (or length) of a vector is denoted by $|a|$ or a. It is a scalar and must be non-negative. Any vector whose length is 1 is called a unit vector; unit vectors will usually be denoted by e.

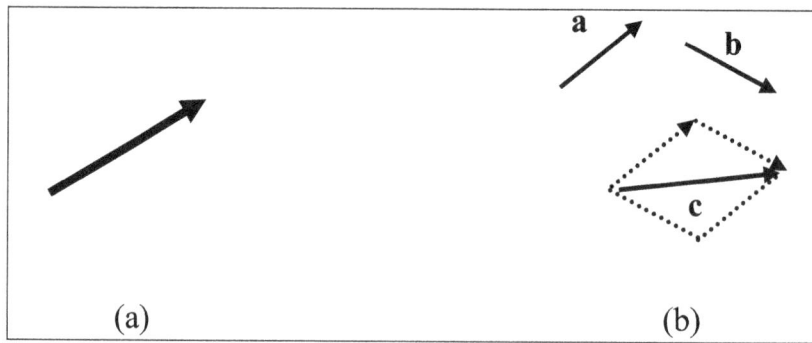

(a) a vector; (b) addition of vectors.

Vector Algebra

The operations of addition, subtraction and multiplication familiar in the algebra of numbers (or scalars) can be extended to an algebra of vectors.

The following definitions and properties fundamentally define the vector:

1. Sum of Vectors:

The addition of vectors a and b is a vector c formed by placing the initial point of b on the terminal point of a and then joining the initial point of a to the terminal point of b. The sum is written c = a + b. This definition is called the parallelogram law for vector addition because, in a geometrical interpretation of vector addition, c is the diagonal of a parallelogram formed by the two vectors a and b, figure. The following properties hold for vector addition:

$$a + b = b + a \qquad \text{... commutative law}$$
$$a + (b+c) = (a + b) + c \qquad \text{... associative law}$$

2. The Negative Vector:

For each vector a there exists a negative vector. This vector has direction opposite to that of vector a but has the same magnitude; it is denoted by $-$ a. A geometrical interpretation of the negative vector is shown in figure.

3. Subtraction of Vectors and the Zero Vector:

The subtraction of two vectors a and b is defined by a $-$ b = a + ($-$b), figure. If a = b then a $-$ b is defined as the zero vector (or null vector) and is represented by the symbol o. It has zero magnitude and unspecified direction. A proper vector is any vector other than the null vector. Thus the following properties hold:

$$a + 0 = a$$
$$a + (-a) = 0$$

4. Scalar Multiplication:

The product of a vector by a scalar α is a vector αa with magnitude α times the magnitude of a and with direction the same as or opposite to that of a, according as α is positive or negative. If

$\alpha = 0$, α a is the null vector. The following properties hold for scalar multiplication:

$(\alpha + \beta)a = \alpha a + \beta a$ *... distributive law, over addition of scalars*

$\alpha(a + b) = \alpha a + \alpha b$ *... distributive law, over addition of vectors*

$a(\beta a) = (\alpha\beta)a$ *... associative law for scalar multiplication*

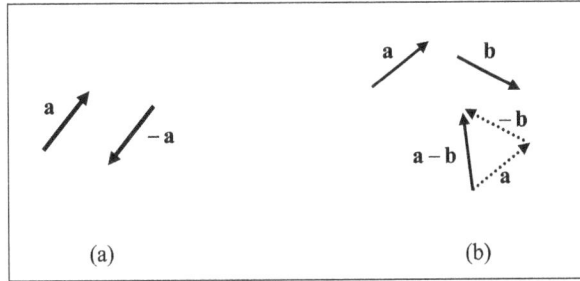

(a)   (b)

(a) Negative of a vector; (b) Subtraction of vectors.

Note that when two vectors a and b are equal, they have the same direction and magnitude, regardless of the position of their initial points. Thus $a = b$ in Fig. A particular position in space is not assigned here to a vector – it just has a magnitude and a direction. Such vectors are called free, to distinguish them from certain special vectors to which a particular position in space is actually assigned.

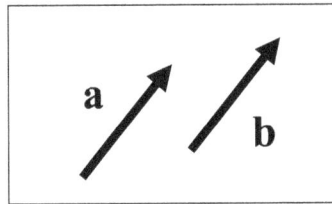

Equal vectors.

The vector as something with "magnitude and direction" and defined by the above rules is an element of one case of the mathematical structure, the vector space.

The Dot Product

The dot product of two vectors a and b (also called the scalar product) is denoted by a · b. It is a scalar defined by,

$$a \cdot b = |a||b| \cos \theta .$$

θ here is the angle between the vectors when their initial points coincide and is restricted to the range $0 \le \theta \le \pi$, in figure below.

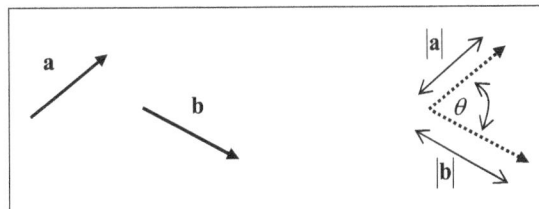

The dot product.

An important property of the dot product is that if for two (proper) vectors a and b, the relation $a \cdot b = 0$, then a and b are perpendicular. The two vectors are said to be orthogonal. Also, $a \cdot a = |a||a|\cos(0)$, so that the length of a vector is $|a| = \sqrt{a \cdot a}$.

Another important property is that the projection of a vector u along the direction of a unit vector e is given by u·e. This can be interpreted geometrically as in figure.

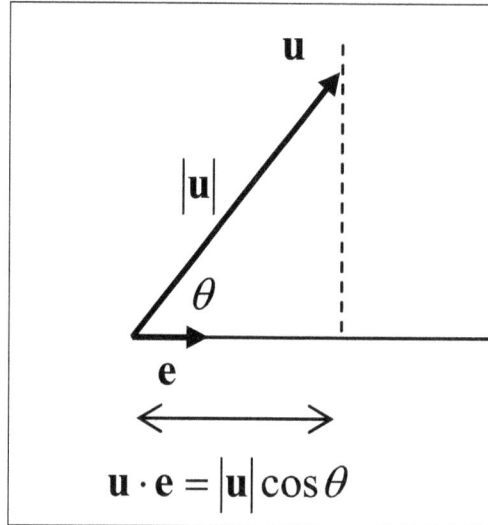

The projection of a vector along the direction of a unit vector.

It follows that any vector u can be decomposed into a component parallel to a (unit) vector e and another component perpendicular to e, according to:

$$u = (u \cdot e)e + [u - (u \cdot e)e]$$

The dot product possesses the following properties (which can be proved using the above definition):

$(1)\, a \cdot b = b \cdot a$ (commutative)

$(2)\, a \cdot (b + c) = a \cdot b + a \cdot c$ (distributive)

$(3)\, \alpha(a \cdot b) = a \cdot (\alpha b)$

$(4)\, a \cdot a \geq 0$; and $a \cdot a = 0$ if and only if $a = o$

Cross Product

The cross product of two vectors a and b (also called the vector product) is denoted by $a \times b$. It is a vector with magnitude:

$$|a \times b| = |a||b|\sin\theta$$

with θ defined as for the dot product. It can be seen from the figure that the magnitude of $a \times b$ is equivalent to the area of the parallelogram determined by the two vectors a and b.

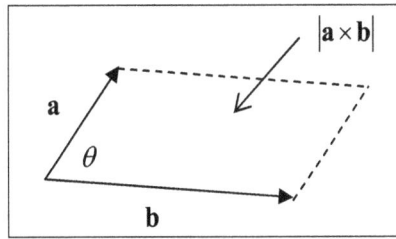

The magnitude of the cross product.

The direction of this new vector is perpendicular to both a and b. Whether $a \times b$ points "up" or "down" is determined from the fact that the three vectors a, b and $a \times b$ form a right handed system. This means that if the thumb of the right hand is pointed in the direction of $a \times b$, and the open hand is directed in the direction of a , then the curling of the fingers of the right hand so that it closes should move the fingers through the angle θ, $0 \leq \theta \leq \pi$, bringing them to b. Some examples are shown in figure below.

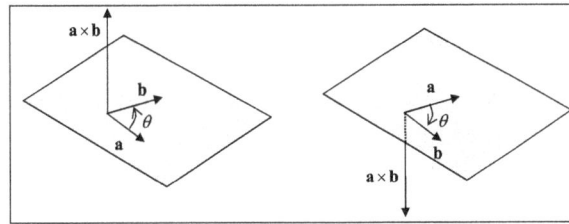

Examples of the cross product.

The cross product possesses the following properties (which can be proved using the above definition):

(1) $a \times b = -a \times b$ (not commutative)

(2) $a \times (b + c) = a \times b + a \times c$ (distributive)

(3) $\alpha(a \times b) = a(\alpha b)$

(4) $a \times b = o$ if and only if a and b $(\neq o)$ are parallel ("linearly dependent")

Triple Scalar Product

The triple scalar product, or box product, of three vectors u, v, w is defined by:

$$\boxed{(u \times v) \cdot w = (v \times w) \cdot u = (w \times u) \cdot v}$$ Triple Scalar Product

Its importance lies in the fact that, if the three vectors form a right-handed triad, then the volume V of a parallelepiped spanned by the three vectors is equal to the box product.

To see this, let e be a unit vector in the direction of $u \times v$, figure. Then the projection of w on $u \times v$ is $h = w \cdot e$, and

$$w \cdot (u \times v) = w \cdot (|u \times v| e)$$
$$= |u \times v| h$$
$$= V$$

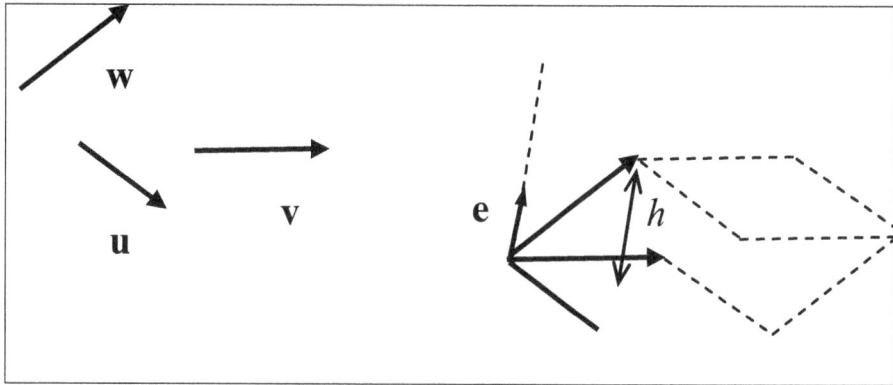

The triple scalar product.

Note: if the three vectors do not form a right handed triad, then the triple scalar product yields the negative of the volume. For example, using the vectors above, $(w \times v) \cdot u = -V$.

Vectors and Points

Vectors are objects which have magnitude and direction, but they do not have any specific location in space. On the other hand, a point has a certain position in space, and the only characteristic that distinguishes one point from another is its position. Points cannot be "added" together like vectors. On the other hand, a vector v can be added to a point p to give a new point $q, q = v + p$, figure. Similarly, the "difference" between two points gives a vector, $q - p = v$. Note that the notion of point as defined here is slightly different to the familiar point in space with axes and origin – the concept of origin is not necessary for these points and their simple operations with vectors.

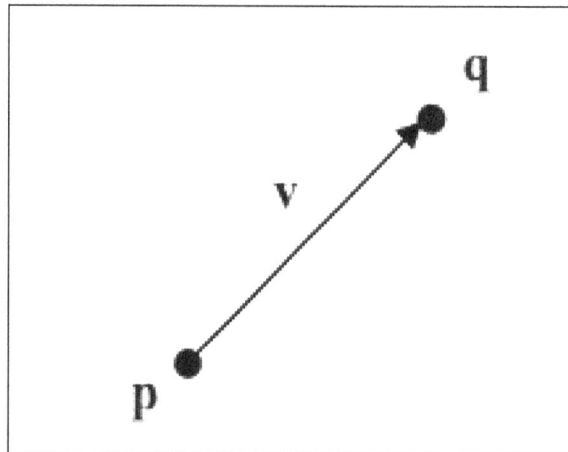

Adding vectors to points.

COMPLEX NUMBERS

A complex number can be represented by an expression of the form $a + bi$, where a and b are real numbers and i is a symbol with the property that $i^2 = -1$. The complex number $a + bi$ can also be represented by the ordered pair (a, b) and plotted as a point in a plane (called the Argand plane) as in figure. Thus, the complex number $i = 0 + 1 \cdot i$ is identified with the point $(0, 1)$.

The real part of the complex number $a + bi$ is the real number a and the imaginary part is the real number b. Thus, the real part of $4 - 3i$ is 4 and the imaginary part is -3.

Two complex numbers $a + bi$ and $c + di$ are equal if $a = c$ and $b = d$, that is, their real parts are equal and their imaginary parts are equal. In the Argand plane the horizontal axis is called the real axis and the vertical axis is called the imaginary axis. The sum and difference of two complex numbers are defined by adding or subtracting their real parts and their imaginary parts:

$$(a + bi) + (c + di) = (a + c) + (b + d)i$$
$$(a + bi) - (c + di) = (a - c) + (b - d)i$$

For instance,

$$(1 - i) + (4 + 7i) = (1 + 4) + (-1 + 7i) = 5 + 6i$$

The product of complex numbers is defined so that the usual commutative and distributive laws hold:

$$(a + bi) + (c + di) = a\,(c + di) + (bi)\,(c + di)$$
$$= ac + adi + bci + bdi^2$$

Since $i^2 = -1$, this becomes:

$$(a + bi) + (c + di) = (ac - bd) + (ad + bc)i$$

Example:

$$(-1 + 3i)\,(2 - 5i) = (-1)\,(2 - 5i) + 3i(2 - 5i)$$
$$= -2 + 5i + 6i - 15(-1) = 13 + 11i$$

Division of complex numbers is much like rationalizing the denominator of a rational expression. For the complex number $z = a + bi$, we define its complex conjugate to be $\bar{z} = a + bi$. To find the quotient of two complex numbers we multiply numerator and denominator by the complex conjugate of the denominator.

Example: Express the number $\dfrac{-1 + 3i}{2 + 5i}$ in the form $a + bi$.

Solution: We multiply numerator and denominator by the complex conjugate of $2 + 5i$, namely $2 - 5i$, and we take advantage of the result of example:

$$\frac{-1 + 3i}{2 + 5i} = \frac{-1 + 3i}{2 + 5i} \cdot \frac{2 - 5i}{2 - 5i} = \frac{13 + 11i}{2^2 + 5^2} = \frac{13}{29} + \frac{11}{29}i$$

The geometric interpretation of the complex conjugate is shown in figure, is the reflection of in the real axis. We list some of the properties of the complex conjugate in the following box.

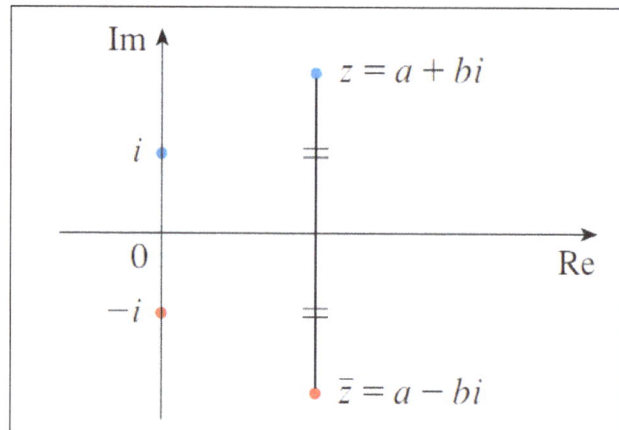

Properties of Conjugates

$$\overline{z + w} = \overline{z} + \overline{w} \qquad \overline{zw} = \overline{z}\ \overline{w} \qquad \overline{z^n} = \overline{z}^n$$

The modulus, or absolute value, $|z|$ of a complex number $z = a + bi$ is its distance from the origin. From figure we see that if $z = a + bi$, then:

$$|z| = \sqrt{a^2 + b^2}$$

Notice that,

$$z\overline{z} = (a + bi)(a - bi) = a^2 + abi - abi - b^2i^2 = a^2 + b^2$$

and so,

$$z\overline{z} = |z|^2$$

This explains why the division procedure in example works in general:

$$\frac{z}{w} = \frac{z\overline{w}}{w\overline{w}} = \frac{z\overline{w}}{|w|^2}$$

Since $i^2 = -1$, we can think of as a square root of -1. But notice that we also have $(-i)^2 = i^2 = -1$ and so $-i$ is also a square root of -1. We say that i is the principal square root of -1 and write $\sqrt{-1} = i$. In general, if c is any positive number, we write:

$$\sqrt{-c} = \sqrt{c}\ i$$

With this convention, the usual derivation and formula for the roots of the quadratic equation $ax^2 + bx + c = 0$ are valid even when $b^2 - a\,4ac < 0$;

$$x = \frac{-b \pm \sqrt{b^2 - 4ac}}{2a}$$

Example: Find the roots of the equation $x^2 + x + 1 = 0$.

Solution: Using the quadratic formula, we have,

$$x = \frac{-1 \pm \sqrt{1^2 - 4 \cdot 1}}{2} = \frac{-1 \pm \sqrt{-3}}{2} = \frac{-1 \pm \sqrt{3}\,i}{2}$$

We observe that the solutions of the equation in Example are complex conjugates of each other. In general, the solutions of any quadratic equation $ax^2 + bx + c = 0$ with real coefficients a, b, and c are always complex conjugates. (If is real, $\bar{z} = z$, so z is its own conjugate.)

We have seen that if we allow complex numbers as solutions, then every quadratic equation has a solution. More generally, it is true that every polynomial equation,

$$a_n x^n + a_{n-1} x^{n-1} + \cdots + a_1 x + a_0 = 0$$

of degree at least one has a solution among the complex numbers. This fact is known as the Fundamental Theorem of Algebra and was proved by Gauss.

Polar Form

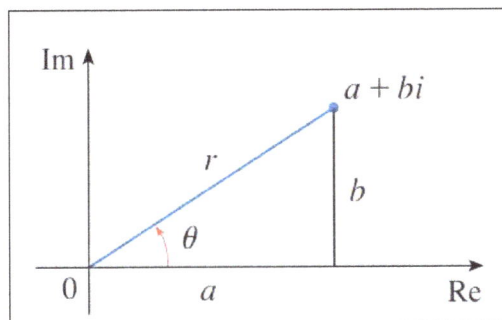

We know that any complex number $z = a + bi$ can be considered as a point (a, b) and that any such point can be represented by polar coordinates (r, θ) with $r \geq 0$. In fact,

$$a = r \cos \theta \qquad\qquad b = r \sin \theta$$

as in figure. Therefore, we have:

$$z = a + bi = (r \cos \theta) + (r \sin \theta)i$$

Thus, we can write any complex number in the form:

$$z = r\,(\cos \theta + i \sin \theta)$$

Where $r = |z| = \sqrt{a^2 + b^2}$ and $\tan \theta = \dfrac{b}{a}$

The angle θ is called the argument of z and we write $\theta = \arg(z)$. Note that $\arg(z)$ is not unique; any two arguments of z differ by an integer multiple of 2π.

Basic Operations

We add, subtract, multiply and divide complex numbers much as we would expect. We add and subtract complex numbers by adding their real and imaginary parts:

$$(a + bi) + (c + di) = (a + c) + (b + d)i,$$
$$(a + bi) - (c + di) = (a - c) + (b - d)i.$$

We can multiply complex numbers by expanding the brackets in the usual fashion and using $i^2 = -1$.

$$(a + bi)(c + di) = ac + bci + adi + bdi^2 = (ac - bd) + (ad + bc)i.$$

To divide complex numbers, we note firstly that $(c + di)(c - di) = c^2 + d^2$ is real. So

$$\frac{a+bi}{c+di} = \frac{a+bi}{c+di} \times \frac{c-di}{c-di} = \left(\frac{ac+bd}{c^2+d^2}\right) + \left(\frac{bc-ad}{c^2+d^2}\right)i.$$

The number $c - di$, which we just used, as relating to $c + di$, has a special name and some useful properties.

Let $z = a + bi$. The conjugate of z is the number $a - bi$, and this is denoted as $\bar{z}$ (or in some books as $z*$).

Note from equation that when the real quadratic equation $ax^2 + bx + c = 0$ has complex roots, then these roots are conjugates of each other. Generally, if the polynomial $a_n z^n + a_{n-1} z^{n-1} + \cdots + a_0 = 0$, where the a_i are real, has a root z_0, then the conjugate $\bar{z}_0$ is also a root.

Example: Find all those z that satisfy $z^2 = i$.

Suppose that $z^2 = i$ and $z = a + bi$, where a and b are real. Then:

$$i = (a + bi)^2 = (a^2 - b^2) + 2abi.$$

Comparing the real and imaginary parts, we know that:

$$a^2 - b^2 = 0 \text{ and } 2ab = 1.$$

So $b = 1/2a$ from the second equation, and substituting for b into the first equation gives $a^4 = 1/4$, which has real solutions $a = 1/\sqrt{2}$ or $a = -1/\sqrt{2}$.

So the two z which satisfy $z^2 = i$, i.e. the two square roots of i, are:

$$\frac{1+i}{\sqrt{2}} \text{ and } \frac{-1-i}{\sqrt{2}}.$$

Example: Use the quadratic formula to find the two solutions of-

$$z^2 - (3+i)\,z\, + \,(2+i) = 0.$$

We see that $a = 1$, $b = -3 - i$, and $c = 2 + i$. So,

$$b^2 - 4ac = (-3-i)^2 - 4 \times 1 \times (2+i) = 9\ -1 + 6i - 8 - 4i = 2i.$$

Knowing $\sqrt{i} = \pm(1+i)/\sqrt{2}$, from the previous problem, we have:

$$
\begin{aligned}
x &= \frac{-b \,\pm\, \sqrt{b^2\, -\, 4ac}}{2a} \\
&= \frac{(3+i) \pm \sqrt{2i}}{2} \\
&= \frac{(3+i)\, \pm \sqrt{2}\sqrt{i}}{2} \\
&= \frac{(3+i) \pm (1+i)}{2} \\
&= \frac{4+2i}{2} \ \text{ or } \ \frac{2}{2} \\
&= 2 + i \text{ or } 1
\end{aligned}
$$

Argand Diagram

The real numbers are often represented on the real line which increase as we move from left to right.

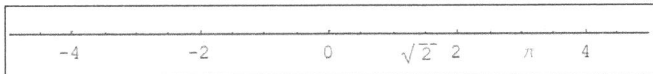

The Real Line.

The complex numbers, having two components, their real and imaginary parts, can be represented as a plane; indeed, C is sometimes referred to as the complex plane, but more commonly, when we represent C in this manner, we call it an Argand diagram. The point $(a,\ b)$ represents the complex number a + bi so that the x-axis contains all the real numbers, and so is termed the real axis, and the y-axis contains all those complex numbers which are purely imaginary (i.e. have no real part), and so is referred to as the imaginary axis.

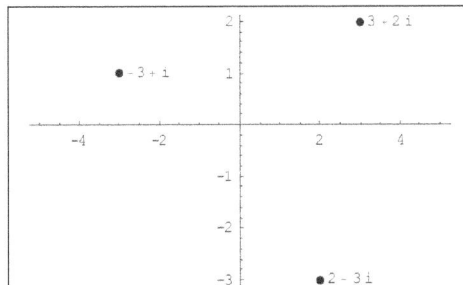

The Argand Diagram.

We can think of $z_0 = a + bi$ as a point in an Argand diagram, but it can often be useful to think of it as a vector as well. Adding z_0 to another complex number translates that number by the vector $\begin{pmatrix} a \\ b \end{pmatrix}$. That is the map $z \mapsto z + z_0$ represents a translation a units to the right and b units up in the complex plane. Note that the conjugate $\bar{z}$ of a point z is its mirror image in the real axis. So, $z \mapsto \bar{z}$ represents reflection in the real axis.

The number r is called the modulus of z and is written $|z|$. The number θ is called the argument of z and is written arg z. If $z = x + iy$ then:

$$|z| = \sqrt{x^2 + y^2} \quad \text{and} \sin \arg z = \frac{y}{\sqrt{x^2 + y^2}}, \cos \arg z = \frac{x}{\sqrt{x^2 + y^2}}.$$

Note that the argument of o is undefined.

Note that arg z is defined only up to multiples of 2π. For example, the argument of $1 + i$ could be $\pi / 4$ or $9\pi / 4$ or $-7\pi / 4$ etc. For simplicity, in this part we shall give all arguments in the range $0 \le \theta < 2\pi$, so that $\pi / 4$ would be the preferred choice here.

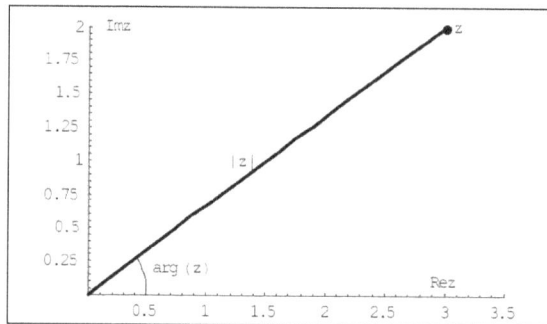

A Complex Number's Cartesian and Polar Co-ordinates.

Let $z, w \in C$. Then:

$$|zw| = |z||w|;$$
$$z\bar{z} = |z|^2;$$
$$\overline{z + w} = \bar{z} + w;$$
$$\overline{zw} = \bar{z}\ \bar{w};$$
$$|z + w| \le |z| + |w|;$$

$$|z / w| = |z| / |w| \quad if\ w \ne 0;$$
$$|\bar{z}| = |z|;$$
$$\overline{z - w} = \bar{z} - \bar{w};$$
$$\overline{z / w} = \bar{z} / \bar{w} \quad if\ w \ne 0;$$
$$||z| - |w|| \le |z - w|;$$

- and up to multiples of 2π then the following equations also hold:

- $\arg(zw) = \arg z + \arg w$ if $z, w \ne 0,$

- $\arg(z / w) = arg\ z - \arg w$ if $z, w \ne 0,$

- $\arg \bar{z} = -\arg z$ if $z \ne 0,$

Proof: $|zw| = |z||w|$. Let $z = a + bi$ and $w = c + di$. Then $zw = (ac - bd) + (bc + ad)i$ so that

$$|zw| = \sqrt{(ac - bd)^2 + (bc + ad)^2}$$
$$= \sqrt{a^2c^2 + b^2d^2 + b^2c^2 + a^2d^2}$$
$$= \sqrt{(a^2 + b^2)(c^2 + d^2)}$$
$$= \sqrt{a^2 + b^2}\sqrt{c^2 + d^2} = |z||w|.$$

Proof: $\arg(zw) = \arg z + \arg w$. Let $z = r(\cos\theta + i\sin\theta)$ and $w = R(\cos\Theta + i\sin\Theta)$. Then:

$$zw = rR(\cos\theta + i\sin\theta)(\cos\Theta + i\sin\Theta)$$
$$= rR((\cos\theta\cos\Theta - \sin\theta\sin\Theta) + i(\sin\theta\cos\Theta + \cos\theta\sin\Theta))$$
$$= rR(\cos(\theta + \Theta) + i\sin(\theta + \Theta)).$$

We can read off that $|zw| = rR = |z||w|$ which is a second proof of the previous part and also that:

$$\arg(zw) = \theta + \Theta = \arg z + \arg w \text{ up to multiples of } 2\pi.$$

Proof: $\overline{zw} = \overline{z}\,\overline{w}$. Let $z = a + bi$ and $w = c + di$. Then,

$$\overline{zw} = \overline{(ac - bd) + (bc + ad)i}$$
$$= (ac - bd) - (bc + ad)i$$
$$= (a - bi)(c - di) = \overline{z}\,\overline{w}$$

Proof: (Triangle Inequality) $|z + w| \le |z| + |w|$

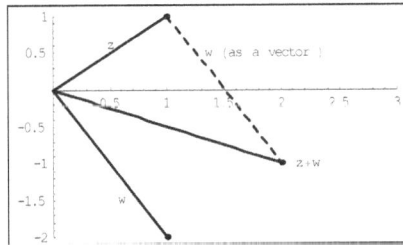

A diagrammatic proof of the Triangle Inequality.

Note that the shortest distance between 0 and $z + w$ is the modulus of $z + w$. This is shorter in length than the path which goes from 0 to z to $z + w$. The total length of this second path is $|z| + |w|$.

For an algebraic proof, note that for any complex number $z + \overline{z} = 2\,\text{Re}\,z$ and $\text{Re}\,z \le |z|$. So for $z, w \in C$,

$$\frac{z\overline{w} + \overline{z}w}{2} = \text{Re}(z\overline{w}) \le |z\overline{w}| = |z||\overline{w}| = |z||w|.$$

Then,

$$|z + w|^2 = (z + w)\overline{(z + w)}$$
$$= (z + w)\ (\overline{z} + \overline{w})$$
$$= z\overline{z} + z\overline{w} + \overline{z}w + w\overline{w}$$
$$\le |z|^2 + 2\ |z|\ |w| + |w|^2 = \big(|z| + |w|\big)^2,$$

to give the required result.

Corollary: The complex roots of a real polynomial come in pairs. That is, if z_0 satisfies the polynomial equation $a_k z^k + a_{k-1} z^{k-1} + \cdots + a_0 = 0$, where each ai is real, then $\overline{z}_0$ is also a root.

Proof: Note from the algebraic properties of the conjugate function, proven in the previous proposition, that:

$$a_k \left(\overline{z}_0\right)^k + a_{k-1}\left(\overline{z}_0\right)^{k-1} + \cdots + a_1 \overline{z}_0 + a_0 = a_k \overline{\left(z_0\right)}^k + a_{k-1}\overline{\left(z_0\right)}^{k-1} + \cdots + a_1 \overline{z}_0 + a_0$$
$$= \overline{a_k \left(z_0\right)}^k + \overline{a_{k-1}\left(z_0\right)}^{k-1} + \cdots + \overline{a_1 z_0} + \overline{a_0}\ \left[\text{the } a_i \text{ are real}\right]$$
$$= \overline{a_k\ \left(z_0\right)^k + a_{k-1}\ \left(z_0\right)^{k-1} + \cdots + a_0}$$
$$= \overline{0}\ \left[\text{as } z_0 \text{ is a root}\right]$$
$$= 0.$$

Roots of Unity

Consider the complex number:

$$z_0 = \cos \theta + i \sin \theta$$

where $0 \le \theta < 2\pi$. The modulus of z_0 is 1, and the argument of z_0 is θ.

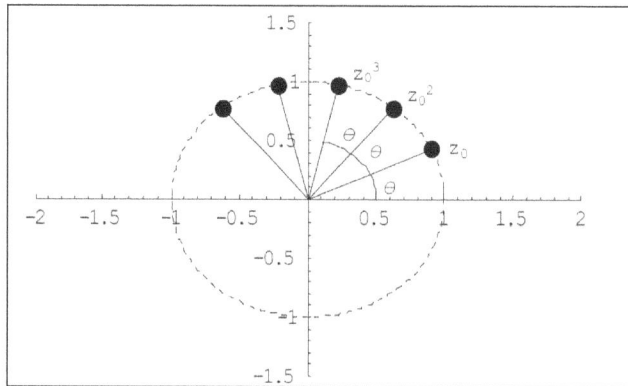

Powers of z_0.

In Proposition we proved for $z, w \ne 0$ that,

$$|zw| = |z|\,|w| \ \text{ and arg } (zw) = \text{arg } z + \text{arg } w,$$

up to multiples of 2π. So for any integer n, and any $z \neq 0$, we have that,

$$|z^n| = |z|^n \text{ and } \arg\left(z^n\right) = n \arg z.$$

So the modulus of $(z_0)^n$ is 1 and the argument of $(z_0)^n$ is $n\theta$, or putting this another way.

Theorem: (de moivre's5 theorem) For a real number θ and integer n we have that,

$$\cos n\theta + i\sin n\theta = \left(\cos \theta + i\sin \theta\right)^n$$

Problem Let n be a natural number. Find all those complex z such that $z^n = 1$.

We know from the Fundamental Theorem of Algebra that there are (counting repetitions) n solutions: these are known as the nth roots of unity.

Let's first solve $z^n = 1$ directly for $n = 2, 3, 4$.

When $n = 2$ we have,

$$0 = z^2 - 1 = (z-1)(z+1)$$

and so the square roots of 1 are ±1.

When $n = 3$ we can factorise as follows:

$$0 = z^3 - 1 = (z-1)\left(z^2 + z + 1\right).$$

So 1 is a root and completing the square we see,

$$0 = z^2 + z + 1 = \left(z + \frac{1}{2}\right)^2 + \frac{3}{4}$$

which has roots,

$$-\frac{1}{2} \pm \frac{\sqrt{3}}{2} i.$$

So the cube roots of 1 are 1, $-1/2 + \sqrt{3}i/2$ and $-1/2 - \sqrt{3}i/2$.

When $n = 4$ we can factorise as follows:

$$0 = z^4 - 1 = \left(z^2 - 1\right)\left(z^2 + 1\right) = (z-1)(z+1)(z-i)(z+i),$$

so that the fourth roots of 1 are 1, -1, i and $-i$.

Plotting these roots on Argand diagrams we can see a pattern developing.

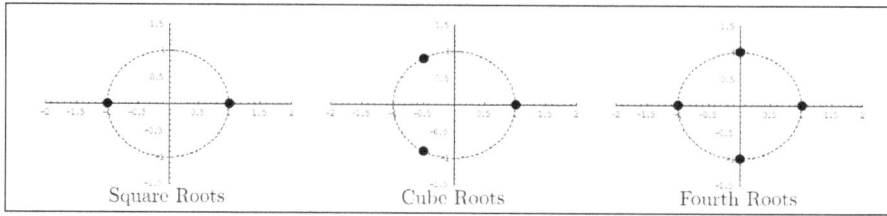

Square Roots   Cube Roots   Fourth Roots

Returning to the general case, suppose that $z = r\left(\cos\theta + i\sin\theta\right)$ and satisfies $z^n = 1$. Then by the observations preceding De Moivre's Theorem z^n has modulus r^n and has argument $n\theta$. As 1 has modulus 1 and argument 0, we can compare their moduli to find rn = 1 giving r = 1. Comparing arguments, we see $n\theta = 0$ up to multiples of 2π. That is $n\theta = 2k\pi$ for some integer k, giving $\theta = 2k\pi / n$. So the roots of $z^n = 1$ are:

$$z = \cos\left(\frac{2k\pi}{n}\right) + i\sin\left(\frac{2k\pi}{n}\right) \quad \text{where k is an integer}$$

At first glance there seem to be an infinite number of roots but note as cos and sin have period 2π then these z repeat with period   . So the nth roots of unity are:

$$z = \cos\left(\frac{2k\pi}{n}\right) + i\sin\left(\frac{2k\pi}{n}\right) \quad where \ k = 0, \ 1, \ 2,...,n-1.$$

Plotted on an Argand diagram, the *nth* roots of unity form a regular ⁿ-gon inscribed within the unit circle with a vertex at 1.

Differential and Integral Calculus

The subfield of calculus which deals with the study of the rates at which quantities change is known as differential calculus. Integral calculus studies the notion and application of integrals and focuses on total size or value, such as lengths, areas, and volumes. The topics elaborated in this chapter will help in gaining a better perspective about these branches of mathematics.

DIFFERENTIAL CALCULUS

A branch of mathematics dealing with the concepts of derivative and differential and the manner of using them in the study of functions. The development of differential calculus is closely connected with that of integral calculus. Indissoluble is also their content. Together they form the base of mathematical analysis, which is extremely important in the natural sciences and in technology. The introduction of variable magnitudes into mathematics by R. Descartes was the principal factor in the creation of differential calculus. Differential and integral calculus were created, in general terms, by I. Newton and G. Leibniz towards the end of the 17th century, but their justification by the concept of limit was only developed in the work of A.L. Cauchy in the early 19th century. The creation of differential and integral calculus initiated a period of rapid development in mathematics and in related applied disciplines. Differential calculus is usually understood to mean classical differential calculus, which deals with real-valued functions of one or more real variables, but its modern definition may also include differential calculus in abstract spaces. Differential calculus is based on the concepts of real number; function; limit and continuity — highly important mathematical concepts, which were formulated and assigned their modern content during the development of mathematical analysis and during studies of its foundations. The central concepts of differential calculus — the derivative and the differential — and the apparatus developed in this connection furnish tools for the study of functions which locally look like linear functions or polynomials, and it is in fact such functions which are of interest, more than other functions, in applications.

Let a function $y = f(x)$ be defined in some neighbourhood of a point x_0. Let $\Delta x \neq 0$ denote the increment of the argument and let $\Delta y = f(x_0 + \Delta x) - f(x_0)$ denote the corresponding increment of the value of the function. If there exists a (finite or infinite) limit.

$$\lim_{\Delta x \to 0} \frac{\Delta y}{\Delta x},$$

then this limit is said to be the derivative of the function f at x_0 it is denoted by $f'(x_0), d\,f(x_0)\,/\,dx, y', y'x, dy\,/\,dx$.

Thus, by definition,

$$f'(x_0) = \lim_{\Delta x \to 0} \frac{\Delta y}{\Delta x} = \lim_{\Delta x \to 0} \frac{f(x_0 + \Delta x) - f(x_0)}{\Delta x}.$$

The operation of calculating the derivative is called differentiation. If $f'(x_0)$ is finite, the function f is called differentiable at the point x_0. A function which is differentiable at each point of some interval is called differentiable in the interval.

Geometric Interpretation of the Derivative

Let C be the plane curve defined in an orthogonal coordinate system by the equation $y = f(x)$ where f is defined and is continuous in some interval J; let $M(x_0, y_0)$ be a fixed point on C, let $P(x, y)$ $(x \in J)$ be an arbitrary point of the curve C and let MP be the secant. An oriented straight line MT (T a variable point with abscissa $x_0 + \Delta x$) is called the tangent to the curve C at the point M if the angle ϕ between the secant MP and the oriented straight line tends to zero as $x \to x_0$ (in other words, as the point $P \in C$ arbitrarily tends to the point M). If such a tangent exists, it is unique. Putting $x = x_0 + \Delta x$, $\Delta y = f(x_0 + \Delta x) - f(x_0)$, one obtains the equation $\tan \quad = \Delta y / \Delta x$ for the angle β between MP and the positive direction of the x-axis.

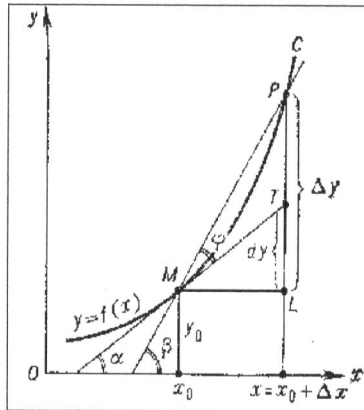

The curve C has a tangent at the point M if and only if $\lim_{\Delta x \to 0} \Delta y / \Delta x$ exists, i.e. if $f'(x_0)$ exists. The equation $\tan \alpha = f'(x_0)$ is valid for the angle α between the tangent and the positive direction of the x-axis. If $f'(x_0)$ is finite, the tangent forms an acute angle with the positive x-axis, i.e. $-\pi / 2 < \alpha < \pi / 2$; if $f'(x_0) = \infty$, the tangent forms a right angle with that axis.

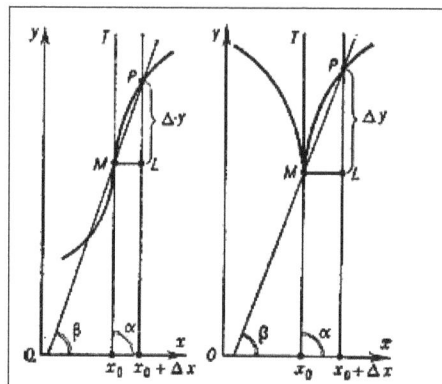

Thus, the derivative of a continuous function f at a point x_0 is identical to the slope $\tan \alpha$ of the tangent to the curve defined by the equation $y = f(x)$ at its point with abscissa x_0.

Mechanical Interpretation of the Derivative

Let a point M move in a straight line in accordance with the law $s = f(t)$ During time Δt the point M becomes displaced by $\Delta s = f(t + \Delta t) - f(t)$. The ratio $\Delta s / \Delta t$ represents the average velocity v_{av} during the time Δt. If the motion is non-uniform, v_{av} is not constant. The instantaneous velocity at the moment t is the limit of the average velocity as $\Delta t \to 0$, i.e. $v = f'(t)$ (on the assumption that this derivative in fact exists).

Thus, the concept of derivative constitutes the general solution of the problem of constructing tangents to plane curves, and of the problem of calculating the velocity of a rectilinear motion. These two problems served as the main motivation for formulating the concept of derivative.

A function which has a finite derivative at a point x_0 is continuous at this point. A continuous function need not have a finite nor an infinite derivative. There exist continuous functions having no derivative at any point of their domain of definition.

The formulas given below are valid for the derivatives of the fundamental elementary functions at any point of their domain of definition (exceptions are stated):

$\quad$ if $f(x) = C = \text{const}$, then $f'(x) = C' = 0$;

$\quad$ if $f(x) = x$, then $f'(x) = 1$;

$\quad$ $(x^\alpha)' = \alpha x^{\alpha-1}, \alpha = \text{const}$ $(x \neq 0,\ \text{if} \ \alpha \leq 1)$;

$\quad$ $(\alpha^x)' = a^x \ \text{In} \ a, a = \text{const} > 0, a \neq 1$; in particular, $(e^x)' = e^x$;

$\quad$ $(\log_\alpha x)' = (\log_\alpha e) / x = 1 / (x \, \text{In} \, a), a = \text{const} > 0, a \neq 1, (\text{In} \, x)' = 1 / x$;

$\quad$ $(\sin x)' = \cos x$;

$\quad$ $(\cos x)' = -\sin x$;

$\quad$ $(\tan x)' = 1 / \cos^2 x$;

$\quad$ $(\cotan x)' = -1 / \sin^2 x$;

$\quad$ $(\arcsin x)' = 1 / \sqrt{1 - x^2}, x \neq \pm 1$;

$\quad$ $(\arccos x)' = -1 / \sqrt{1 - x^2}, x \neq \pm 1$;

$\quad$ $(\arctan x)' = 1 / (1 + x^2)$;

$\quad$ $(\text{arccot} \, an \, x)' = -1 / (1 + x^2)$;

$\quad$ $(\sinh x)'; = \cosh x$;

$\quad$ $(\cosh x)' = \sinh x$;

$\quad$ $(\tanh x)' = 1 / \cosh^2 x$;

$\quad$ $(\cotanh x)' = -1 / \sinh^2 x$.

The following laws of differentiation are valid:

If two functions u and v are differentiable at a point x_0, then the functions:

$$cu \text{ (where c = const)}, u \pm v \ uv, \frac{u}{v} (v \neq 0)$$

are also differentiable at that point, and

$$(cu)' = cu',$$

$$(u \pm v)' = u' \pm v',$$

$$(uv)' = u'v + uv'$$

$$\left(\frac{u}{v}\right)' = \frac{u'v - uv'}{v^2}.$$

Theorem on the derivative of a composite function: If the function $y = f(u)$ is differentiable at a point u_0 while the function $\phi(x)$ is differentiable at a point x_0, and if $u_0 = \phi(x_0)$, then the composite function $y = f(\phi(x))$ is differentiable at x_0, and $y'_x = f'(u_0) \phi'(x_0)$ or, using another notation, $dy / dx = (dy / du)(du / dx)$.

Theorem on the derivative of the inverse function: If $y = f(x)$ and $x = g(y)$ are two mutually inverse increasing (or decreasing) functions, defined on certain intervals, and if $f'(x_0) \neq 0$ exists (i.e. is not infinite), then at the point $y_0 = f(x_0)$ the derivative $g'(y_0) = 1 / f'(x_0)$ exists, or, in a different notation, $dx / dy = 1 / (dy / dx)$. This theorem may be extended: If the other conditions hold and if also $f'(x_0) = 0$ or $f'(x_0) = \infty$ then, respectively, $g'(y_0) = \infty$ or $g'(y_0) = 0$.

One-sided Derivatives

If at a point x_0 the limit,

$$\lim_{\Delta x \downarrow 0} \frac{\Delta y}{\Delta x}$$

exists, it is called the right-hand derivative of the function $y = f(x)$ at x_0 (in such a case the function need not be defined everywhere in a certain neighbourhood of the point x_0; this requirement may then be restricted to $x \geq x_0$). The left-hand derivative is defined in the same way, as:

$$\lim_{\Delta x \uparrow 0} \frac{\Delta y}{\Delta x}$$

A function f has a derivative at a point x_0 if and only if equal right-hand and left-hand derivatives exist at that point. If the function is continuous, the existence of a right-hand derivative at a point is equivalent to the existence, at the corresponding point of its graph, of a right one-sided semi-tangent with slope equal to the value of this one-sided derivative. Points at which the semi-tangents do not form a straight line are called angular points or cusps.

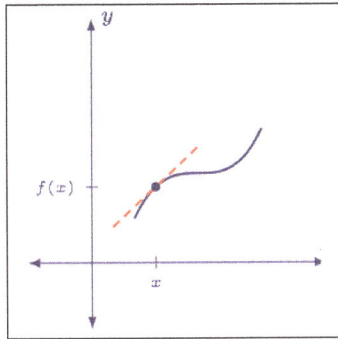

Derivatives of Higher Orders

Let a function $y = f(x)$ have a finite derivative $y' = f'(x)$ at all points of some interval; this derivative is also known as the first derivative, or the derivative of the first order, which, being a function of x, may in its turn have a derivative $y'' = f''(x)$, known as the second derivative, or the derivative of the second order, of the function f, etc. In general, the n-th derivative, or the derivative of order n, is defined by induction by the equation $y^{(n)} = (y^{(n-1)})'$, on the assumption that $y^{(n-1)}$ is defined on some interval. The notations employed along with $y^{(n)}$ are $f^{(n)}$, $d^n f(x) / dx^n$, and, if $n = 2, 3$, also f'', $f''(x)$, f''', $f'''(x)$.

The second derivative has a mechanical interpretation: It is the acceleration $w = d^2 s / dt^2 = f''(t)$ of a point in rectilinear motion according to the law $s = f(t)$.

Differential

Let a function $y = f(x)$ be defined in some neighbourhood of a point x and let there exist a number A such that the increment Δy may be represented as $\Delta y = A \Delta x + w$ with $w / \Delta x \to 0$ as $\Delta x \to 0$. The term Δ in this sum is denoted by the symbol dy or df and is named the differential of the function $f(x)$ (with respect to the variable x) at x. The differential is the principal linear part of increment of the function (its geometrical expression is the segment LT in figure, where MT is the tangent to $y = f(x)$ at the point (x_0, y_0) under consideration).

The function $y = f(x)$ has a differential at x if and only if it has a finite derivative,

$$f'(x) = \lim_{\Delta x \to 0} \frac{\Delta y}{\Delta x} = A$$

at this point. A function for which a differential exists is called differentiable at the point in question. Thus, the differentiability of a function implies the existence of both the differential and the finite derivative, and $dy = df(x) = f'(x) \Delta x$. For the independent variable x one puts $dx = \Delta x$, and one may accordingly write $dy = f'(x) dx$, i.e. the derivative is equal to the ratio of the differentials:

$$f'(x) = \frac{dy}{dx}.$$

The formulas and the rules for computing derivatives lead to corresponding formulas and rules for calculating differentials. In particular, the theorem on the differential of a composite function is valid: If a function $y = f(u)$ is differentiable at a point u_0, while a function $\phi(x)$ is differentiable at a point x_0 and $u_0 = \phi(x_0)$, then the composite function $y = f(\phi(x))$ is differentiable at the point x_0 and $dy = f'(u_0)\, du$, where $du = \phi'(x_0)dx$. The differential of a composite function has exactly the form it would have if the variable u were an independent variable. This property is known the invariance of the form of the differential. However, if u is an independent variable, $du = \Delta u$ is an arbitrary increment, but if u is a function, is the differential of this function which, in general, is not identical with its increment.

Differentials of Higher Orders

The differential dy is also known as the first differential, or differential of the first order. Let $y = f(x)$ have a differential $dy = f'(x)\, dx$ at each point of some interval. Here $dx = \Delta x$ is some number independent of x and one may say, therefore, that $dx = \text{const}$. The differential dy is a function of x alone, and may in turn have a differential, known as the second differential, or the differential of the second order, of f, etc. In general, the n-th differential, or the differential of order n, is defined by induction by the equality $d^n y = d(d^{n-1} y)$, on the assumption that the differential $d^{n-1} y$ is defined on some interval and that the value of dx is identical at all steps. The invariance condition for $d^2 y, d^3 y, \ldots$, is generally not satisfied (with the exception $y = f(u)$ where u is a linear function).

The repeated differential of dy has the form,

$$\delta(dy) = f''(x)\, dx\, \delta x$$

and the value of $\delta(dy)$ for $dx = \delta x$ is the second differential.

Principal Theorems and Applications of Differential Calculus

The fundamental theorems of differential calculus for functions of a single variable are usually considered to include the Rolle theorem, the Legendre theorem (on finite variation), the Cauchy theorem, and the Taylor formula. These theorems underlie the most important applications of differential calculus to the study of properties of functions — such as increasing and decreasing functions, convex and concave graphs, finding the extrema, points of inflection, and the asymptotes of a graph. Differential calculus makes it possible to compute the limits of a function in many cases when this is not feasible by the simplest limit theorems. Differential calculus is extensively applied in many fields of mathematics, in particular in geometry.

Differential Calculus of Functions in Several Variables

For the sake of simplicity the case of functions in two variables (with certain exceptions) is considered below, but all relevant concepts are readily extended to functions in three or more variables. Let a function $z = f(x, y)$ be given in a certain neighbourhood of a point (x_0, y_0) and let the value $y = y_0$ be fixed. $f(x, y_0)$ will then be a function of x alone. If it has a derivative with respect to at x_0, this derivative is called the partial derivative of f with respect to x at (x_0, y_0); it is denoted by $f_x'(x_0, y_0)$, $\delta f(x_0, y_0) / \delta x$, $\delta f / \delta x$, z_x', $\delta z / \delta x$, or $f_x(x_0, y_0)$. Thus, by definition,

$$f_x'(x_0, y_0) = \lim_{\Delta y \to 0} \frac{\Delta_y z}{\Delta y} = \lim_{\Delta y \to 0} \frac{f(x_0, y_0 + \Delta y) - f(x_0, y_0)}{\Delta y},$$

where $\Delta_x z = f(x_0, \Delta x, y_0) - f(x_0, y_0)$ is the partial increment of the function with respect to x (in the general case, $\partial z / \partial x$ must not be regarded as a fraction; $\partial / \partial x$ is the symbol of an operation).

The partial derivative with respect to y is defined in a similar manner:

$$f_x'(x_0, y_0) = \lim_{\Delta y \to 0} \frac{\Delta_y z}{\Delta y} = \lim_{\Delta y \to 0} \frac{f(x_0, y_0 + \Delta y) - f(x_0, y_0)}{\Delta y},$$

where $\Delta_y z$ is the partial increment of the function with respect to y. Other notations include $\partial f(x_0, y_0) / \partial y, \partial f / \partial y, z_y', \partial z / \partial y$ and $fy(x_0, y_0)$. Partial derivatives are calculated according to the rules of differentiation of functions of a single variable (in computing z_x' one assumes $y = \mathrm{const}$ while if z_y' is calculated, one assumes $x = \mathrm{const}$).

The partial differentials of $z = f(x, y)$ at (x_0, y_0) are, respectively,

$$d_x z = f_x'(x_0, y_0)\, dx; \quad d_y z = f_y'(x_0, y_0)\, dy$$

where, as in the case of a single variable, $dx = \Delta x$, $dy = \Delta y$ denote the increments of the independent variables.

The first partial derivatives $\partial z / \partial x = f_x'(x, y)$ and $\partial z / \partial y = f_x'(x, y)$, or the partial derivatives of the first order, are functions of x and y, and may in their turn have partial derivatives with respect to x and y. These are named, with respect to the function $z = f(x, y)$, the partial derivatives of the second order, or second partial derivatives. It is assumed that,

$$\frac{\partial}{\partial x}\left(\frac{\partial z}{\partial x}\right) = \frac{\partial^2 z}{\partial x^2}, \quad \frac{\partial}{\partial x}\left(\frac{\partial z}{\partial x}\right) = \frac{\partial^2 z}{\partial x\, \partial y}$$

$$\frac{\partial}{\partial x}\left(\frac{\partial z}{\partial y}\right) = \frac{\partial^2 z}{\partial y\, \partial x}, \quad \frac{\partial}{\partial y}\left(\frac{\partial z}{\partial y}\right) = \frac{\partial^2 z}{\partial y^2}$$

The following notations are also used instead of $\partial^2 z / \partial x^2$:

$$z_{xx}'', \quad z_{x^2}'', \quad \frac{\partial^2 f(x, y)}{\partial x^2}, \quad \frac{\partial^2 f}{\partial x^2}, \quad f_{xx}''(x, y), \quad f_{x^2}''(x, y)\ f_{xx}(x, y);$$

and instead of $\partial^2 z / \partial x\, \partial y$:

$$z_{xy}'', \quad \frac{\partial^2 f(x, y)}{\partial x \partial y}, \quad \frac{\partial^2 f}{\partial x \partial y}, \quad f_{xy}''(x, y), f_{xy}(x, y),$$

etc. One can introduce in the same manner partial derivatives of the third and higher orders, together with the respective notations: $\partial^n z / \partial x^n$ means that the function z is to be differentiated n times with respect to x; $\partial^n z / \partial x^P \partial y^q$ where $n = P + q$ means that the function z is differentiated P times with respect to n and q times with respect to y. The partial derivatives of second and higher orders obtained by differentiation with respect to different variables are known as mixed partial derivatives.

To each partial derivative corresponds some partial differential, obtained by its multiplication by the differentials of the independent variables taken to the powers equal to the number of differentiations with respect to the respective variable. In this way one obtains the n-th partial differentials, or the partial differentials of order n:

$$\frac{\partial^n z}{\partial x^n}\, dx^n\ ,\quad \frac{\partial^n z}{\partial x^p\, \partial y^q}\, dx^p\, dy^q\ .$$

The following important theorem on derivatives is valid: If, in a certain neighbourhood of a point (x_0, y_0), a function $z = f(x, y)$ has mixed partial derivatives $f''_{xy}(x, y)$ and $f''_{yx}(x, y)$, and if these derivatives are continuous at the point (x_0, y_0), then they coincide at this point.

A function $z = f(x, y)$ is called differentiable at a point (x_0, y_0) with respect to both variables x and y if it is defined in some neighbourhood of this point, and if its total increment,

$$\Delta z = f(x_0 + \Delta x, y_0 + \Delta y) - f(x_0, y_0)$$

may be represented in the form,

$$\Delta z = A\, \Delta x\ B\, \Delta y + w$$

where A and B are certain numbers and $\omega / \rho \to 0$ for $\rho = \sqrt{(\Delta x)^2 + (\Delta y)^2} \to 0$ (provided that the point $(x_0 + \Delta x, y_0 + \Delta y)$ lies in this neighbourhood). In this context, the expression,

$$dz = d\, f\, (x_0, y_0) = A\, \Delta x + B\, \Delta y$$

is called the total differential (of the first order) of f at (x_0, y_0); this is the principal linear part of increment. A function which is differentiable at a point is continuous at that point (the converse proposition is not always true). Moreover, differentiability entails the existence of finite partial derivatives,

$$f'_x(x_0, y_0) = \lim_{\Delta x \to 0} \frac{\Delta x\, z}{\Delta x} = A,\quad f'_y(x_0, y_0) = \lim_{\Delta y \to 0} \frac{\Delta x\, z}{\Delta y} = B\ .$$

Thus, for a function which is differentiable at (x_0, y_0),

$$dz = df\, (x_0, y_0) = f'_x(x_0, y_0)\, \Delta x + f'_y(x_0, y_0)\, \Delta y\ ,$$

or

$$dz = df\, (x_0, y_0) = f'_x(x_0, y_0)\, dx + f'_y(x_0, y_0)\, dy\ ,$$

if, as in the case of a single variable, one puts, for the independent variables, $dx = \Delta x$, $dy = \Delta y$.

The existence of finite partial derivatives does not, in the general case, entail differentiability (unlike in the case of functions in a single variable). The following is a sufficient criterion of the differentiability of a function in two variables: If, in a certain neighbourhood of a point (x_0, y_0), a function f has finite partial derivatives f'_x and f'_y which are continuous at (x_0, y_0), then f is differentiable at this point. Geometrically, the total differential $df(x_0, y_0)$ is the increment of the applicate of the tangent plane to the surface $z = f(x, y)$ at the point (x_0, y_0, z_0), where $z_0 = f(x_0, y_0)$.

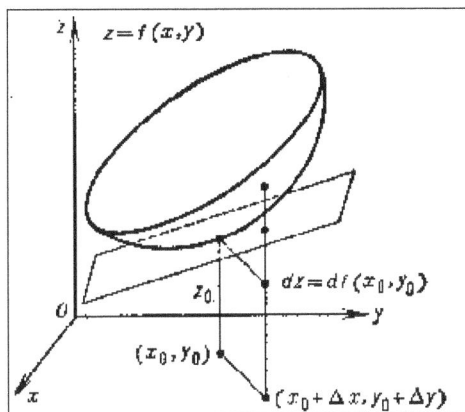

Total differentials of higher orders are, as in the case of functions of one variable, introduced by induction, by the equation,

$$d^n z = d(d^{n-1} z)$$

on the assumption that the differential d^{n-1} is defined in some neighbourhood of the point under consideration, and that equal increments of the arguments d_x, d_y are taken at all steps. Repeated differentials are defined in a similar manner.

Derivatives and Differentials of Composite Functions

Let $w = f(u_1, \ldots, u_m)$ be a function in m variables which is differentiable at each point of an open domain D of the m-dimensional Euclidean space $\mathbf{R}^m$, and let m functions $u_1 = \phi_1(x_1, \ldots, x_n), \ldots, u_m = \phi_m(x_1, \ldots, x_n)$ in n variables be defined in an open domain G of the n-dimensional Euclidean space $\mathbf{R}^n$. Finally, let the point $(u_1, \ldots, u_m)$, corresponding to a point $(x_1, \ldots, x_n) \in G$, be contained in D. The following theorems then hold:

A) If the functions $\phi_1, \ldots, \phi_m$ have finite partial derivatives with respect to $x_1, \ldots, x_n$, the composite function $w = f(u_1, \ldots, u_m)$ in $x_1, \ldots, x_n$ also has finite partial derivatives with respect to $x_1, \ldots, x_n$, and

$$\frac{\partial w}{\partial x_1} = \frac{\partial f}{\partial u_1} \frac{\partial u_1}{\partial x_1} + \ldots + \frac{\partial f}{\partial u_n} \frac{\partial u_n}{\partial x_1}$$

$$\frac{\partial w}{\partial x_n} = \frac{\partial f}{\partial u_1} \frac{\partial u_1}{\partial x_n} + \ldots + \frac{\partial f}{\partial u_n} \frac{\partial u_n}{\partial x_1}$$

B) If the functions $\phi_1, \ldots, \phi_m$ are differentiable with respect to all variables at a point $(x_1, \ldots, x_n) \in G$, then the composite function $w = f(u_1, \ldots, u_m)$ is also differentiable at that point, and

$$dw = \frac{\partial f}{\partial u_1} du_1 + \ldots + \frac{\partial f}{\partial u_n} du_n$$

where $du_1,\ldots,du_m$ are the differentials of the functions $u_1,\ldots,u_m$. Thus, the property of invariance of the first differential also applies to functions in several variables. It does not usually apply to differentials of the second or higher orders.

Differential calculus is also employed in the study of the properties of functions in several variables: finding extrema, the study of functions defined by one or more implicit equations, the theory of surfaces, etc. One of the principal tools for such purposes is the Taylor formula.

The concepts of derivative and differential and their simplest properties, connected with arithmetical operations over functions and superposition of functions, including the property of invariance of the first differential, are extended, practically unchanged, to complex-valued functions in one or more variables, to real-valued and complex-valued vector functions in one or several real variables, and to complex-valued functions and vector functions in one or several complex variables. In functional analysis the ideas of the derivative and the differential are extended to functions of the points in an abstract space.

MEAN VALUE THEOREM

In mathematics, the mean value theorem states, roughly, that for a given planar arc between two endpoints, there is at least one point at which the tangent to the arc is parallel to the secant through its endpoints.

This theorem is used to prove statements about a function on an interval starting from local hypotheses about derivatives at points of the interval.

More precisely, if f is a continuous function on the closed interval $[a,b]$, and differentiable on the open interval (a,b), then there exists a point c in (a,b) such that:

$$f'(c) = \frac{f(b)-f(a)}{b-a}.$$

It is one of the most important results in real analysis.

Formal Statement

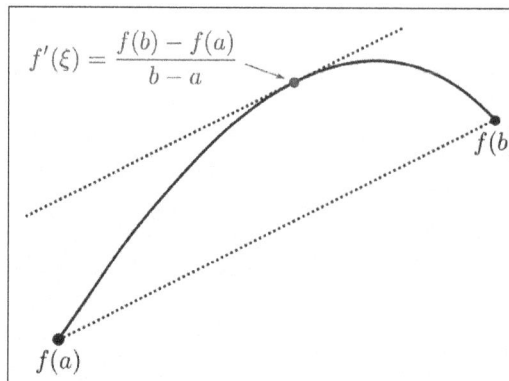

The function f attains the slope of the secant between a and b as the derivative at the point $\xi \in (a,b)$.

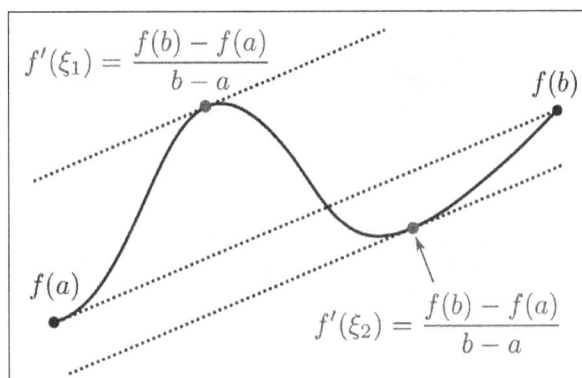

It is also possible that there are multiple tangents parallel to the secant.

Let $f : [a,b] \rightarrow$ be a continuous function on the closed interval $[a,b]$, and differentiable on the open interval (a,b), where $a < b$. Then there exists some c in (a,b) such that:

$$f'(c) = \frac{f(b) - f(a)}{b - a}.$$

The mean value theorem is a generalization of Rolle's theorem, which assumes $f(a) = f(b)$, so that the right-hand side above is zero.

The mean value theorem is still valid in a slightly more general setting. One only needs to assume that $f : [a,b] \rightarrow \mathbb{R}$ is continuous on $[a,b]$, and that for every x in (a,b) the limit:

$$\lim_{h \to 0} \frac{f(x+h) - f(x)}{h}$$

exists as a finite number or equals ∞ or $-\infty$. If finite, that limit equals $f'(x)$. An example where this version of the theorem applies is given by the real-valued cube root function mapping $x \rightarrow x^{\frac{1}{3}}$, whose derivative tends to infinity at the origin.

Note that the theorem, as stated, is false if a differentiable function is complex-valued instead of real-valued. For example, define $f(x) = e^{xi}$ for all real x. Then,

$$f(2\pi) - f(0) = 0 = 0(2\pi - 0)$$

while $f'(x) \neq 0$ for any real x.

These formal statements are also known as Lagrange's Mean Value Theorem.

The expression $\frac{f(b) - f(a)}{b - a}$ gives the slope of the line joining the points $(a, f(a))$ and $(b, f(b))$, which is a chord of the graph of f, while $f'(x)$ gives the slope of the tangent to the curve at the point $(x, f(x))$. Thus the mean value theorem says that given any chord of a smooth curve, we can find a point lying between the end-points of the chord such that the tangent at that point is parallel to the chord. The following proof illustrates this idea.

Define $g(x) = f(x) - rx$, where r is a constant. Since f is continuous on $[a,b]$ and differentiable on (a,b), the same is true for g. We now want to choose r so that g satisfies the conditions of Rolle's theorem. Namely,

$$g(a) = g(b) \Leftrightarrow f(a) - ra = f(b) - rb$$
$$\Leftrightarrow r(b-a) = f(b) - f(a)$$
$$\Leftrightarrow r = \frac{f(b) - f(a)}{b - a}.$$

By Rolle's theorem, since g is differentiable and $g(a) = g(b)$, there is some c in f(a,b)or which $g'(c) = 0$, and it follows from the equality $g(x) = f(x) - rx$ that,

$$g'(x) = f'(x) - r$$
$$g'(c) = 0$$
$$g'(c) = f'(c) - r = 0$$
$$\Rightarrow f'(c) = r = \frac{f(b) - f(a)}{b - a}$$

Simple Application

Assume that f is a continuous, real-valued function, defined on an arbitrary interval I of the real line. If the derivative of f at every interior point of the interval I exists and is zero, then f is constant in the interior.

Proof: Assume the derivative of f at every interior point of the interval I exists and is zero. Let (a, b) be an arbitrary open interval in I. By the mean value theorem, there exists a point c in (a,b) such that:

$$0 = f'(c) = \frac{f(b) - f(a)}{b - a}.$$

This implies that $f(a) = f(b)$. Thus, f is constant on the interior of I and thus is constant on I by continuity.

Remarks:

- Only continuity of f, not differentiability, is needed at the endpoints of the interval I. No hypothesis of continuity needs to be stated if I is an open interval, since the existence of a derivative at a point implies the continuity at this point.

- The differentiability of f can be relaxed to one-sided differentiability.

Cauchy's Mean Value Theorem

Cauchy's mean value theorem, also known as the extended mean value theorem, is a generalization of the mean value theorem. It states: If functions f and g are both continuous on the closed interval $[a, b]$, and differentiable on the open interval (a, b), then there exists some $c \in (a,b)$, such that:

$$(f(b)-f(a))g'(c)=(g(b)-g(a))f'(c).$$

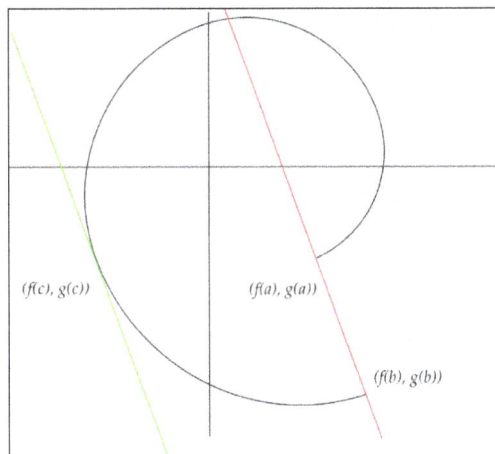

Geometrical meaning of Cauchy's theorem.

Of course, if $g(a) \neq g(b)$ and if $g'(c) \neq 0$, this is equivalent to:

$$\frac{f'(c)}{g'(c)}=\frac{f(b)-f(a)}{g(b)-g(a)}.$$

Geometrically, this means that there is some tangent to the graph of the curve

$$\begin{cases}[a,b]\to \mathbf{R}^2\\ t\mapsto (f(t),g(t))\end{cases}$$

which is parallel to the line defined by the points $(f(a), g(a))$ and $(f(b), g(b))$. However Cauchy's theorem does not claim the existence of such a tangent in all cases where $(f(a), g(a))$ and $(f(b), g(b))$ are distinct points, since it might be satisfied only for some value c with $f'(c) = g'(c) = 0$, in other words a value for which the mentioned curve is stationary; in such points no tangent to the curve is likely to be defined at all. An example of this situation is the curve given by

$$t\mapsto (t^3,1-t^2),$$

which on the interval $[-1, 1]$ goes from the point $(-1, 0)$ to $(1, 0)$, yet never has a horizontal tangent; however it has a stationary point (in fact a cusp) at $t = 0$.

Cauchy's mean value theorem can be used to prove l'Hôpital's rule. The mean value theorem is the special case of Cauchy's mean value theorem when $g(t) = t$.

Proof of Cauchy's Mean Value Theorem

The proof of Cauchy's mean value theorem is based on the same idea as the proof of the mean value theorem.

Suppose $g(a) \neq g(b)$. Define $h(x) = f(x) - rg(x)$, where r is fixed in such a way that $h(a) = h(b)$, namely:

$$h(a) = h(b) \Leftrightarrow f(a) - rg(a) = f(b) - rg(b)$$
$$\Leftrightarrow r(g(b) - g(a)) = f(b) - f(a)$$
$$\Leftrightarrow r = \frac{f(b) - f(a)}{g(b) - g(a)}.$$

Since f and g are continuous on $[a, b]$ and differentiable on (a, b), the same is true for h. All in all, h satisfies the conditions of Rolle's theorem: consequently, there is some c in (a, b) for which $h'(c)$ = 0. Now using the definition of h we have:

$$0 = h'(c) = f'(c) - rg'(c) = f'(c) - \left(\frac{f(b) - f(a)}{g(b) - g(a)} \right) g'(c).$$

Therefore:

$$f'(c) = \frac{f(b) - f(a)}{g(b) - g(a)} g'(c),$$

which implies the result.

If $g(a) = g(b)$, then, applying Rolle's theorem to g, it follows that there exists c in (a, b) for which $g'(c) = 0$. Using this choice of c, Cauchy's mean value theorem (trivially) holds.

Generalization for Determinants

Assume that $f, g,$ and h are differentiable functions on (a,b) that are continuous on $[a,b]$. Define:

$$D(x) = \begin{vmatrix} f(x) & g(x) & h(x) \\ f(a) & g(a) & h(a) \\ f(b) & g(b) & h(b) \end{vmatrix}$$

There exists $c \in (a,b)$ such that $D'(c) 0$.

Notice that,

$$D'(x) = \begin{vmatrix} f'(x) & g'(x) & h'(x) \\ f(a) & g(a) & h(a) \\ f(b) & g(b) & h(b) \end{vmatrix}$$

and if we place $h(x) = 1$, we get Cauchy's mean value theorem. If we place $h(x) = 1$ and $g(x) = x$ we get Lagrange's mean value theorem.

The proof of the generalization is quite simple: each of $D(a)$ and $D(b)$ are determinants with two identical rows, hence $D(a) = D(b) = 0$. The Rolle's theorem implies that there exists $c \in (a,b)$ such that $D'(c) = 0$.

Mean Value Theorem in Several Variables

The mean value theorem generalizes to real functions of multiple variables. The trick is to use parametrization to create a real function of one variable, and then apply the one-variable theorem.

Let G be an open convex subset of $\mathbb{R}^n$, and let $f: G \to \mathbb{R}$ be a differentiable function. Fix points $x, y \in G$, and define $g(t) = f((1-t)x + ty)$. Since g is a differentiable function in one variable, the mean value theorem gives:

$$g(1) - g(0) = g'(c)$$

for some c between 0 and 1. But since $g(1) = f(y)$ and $g(0) = f(x)$, computing $g'(c)$ explicitly we have:

$$f(y) - f(x) = \nabla f\big((1-c)x + cy\big) \cdot (y - x)$$

where ∇ denotes a gradient and $\cdot$ a dot product. Note that this is an exact analog of the theorem in one variable (in the case $n = 1$ this *is* the theorem in one variable). By the Cauchy–Schwarz inequality, the equation gives the estimate:

$$\big|f(y) - f(x)\big| \le \big|\nabla f\big((1-c)x + cy\big)\big|\,\big|y - x\big|.$$

In particular, when the partial derivatives of f are bounded, f is Lipschitz continuous (and therefore uniformly continuous). Note that f is not assumed to be continuously differentiable or continuous on the closure of G. However, in order to use the chain rule to compute g', we really do need to know that f is differentiable on G; the existence of the x and y partial derivatives by itself is not sufficient for the theorem to be true.

As an application of the above, we prove that f is constant if G is open and connected and every partial derivative of f is 0. Pick some point $x_0 \in G$, and let $g(x) = f(x) - f(x_0)$. We want to show $g(x) = 0$ for every $x \in G$. For that, let $E = \{x \in G : g(x) = 0\}$. Then E is closed and nonempty. It is open too: for every $x \in E$,

$$|g(y)| = |g(y) - g(x)| \le (0)\,|y - x| = 0$$

for every y in some neighborhood of x. (Here, it is crucial that x and y are sufficiently close to each other.) Since G is connected, we conclude $E = G$.

The above arguments are made in a coordinate-free manner; hence, they generalize to the case when G is a subset of a Banach space.

Mean Value Theorem for Vector-valued Functions

There is no exact analog of the mean value theorem for vector-valued functions.

In *Principles of Mathematical Analysis,* Rudin gives an inequality which can be applied to many of the same situations to which the mean value theorem is applicable in the one dimensional case:

Theorem: *For a continuous vector-valued function* $\mathbf{f}:[a,b]\to\mathbb{R}^k$ *differentiable on* (a,b), *there exists* $x\in(a,b)$ *such that* $|\mathbf{f}'(x)|\geq\dfrac{1}{b-a}|\mathbf{f}(b)-\mathbf{f}(a)|$.

Jean Dieudonné in his classic treatise *Foundations of Modern Analysis* discards the mean value theorem and replaces it by mean inequality as the proof is not constructive and one cannot find the mean value and in applications one only needs mean inequality. Serge Lang in *Analysis I* uses the mean value theorem, in integral form, as an instant reflex but this use requires the continuity of the derivative. If one uses the Henstock–Kurzweil integral one can have the mean value theorem in integral form without the additional assumption that derivative should be continuous as every derivative is Henstock–Kurzweil integrable. The problem is roughly speaking the following: If $f:U\to\mathbb{R}^m$ is a differentiable function (where $U\subset\mathbb{R}^n$ is open) and if $x+th,\,x,\,h\in\mathbb{R}^n,\,t\in[0,1]$ is the line segment in question (lying inside U), then one can apply the above parametrization procedure to each of the component functions f_i ($i=1,\dots,m$) of f (in the above notation set $y=x+h$). In doing so one finds points $x+t_ih$ on the line segment satisfying.

$$f_i(x+h)-f_i(x)=\nabla f_i(x+t_ih)\cdot h.$$

But generally there will not be a *single* point $x+t^*h$ on the line segment satisfying:

$$f_i(x+h)-f_i(x)=\nabla f_i(x+t^*h)\cdot h.$$

for all i *simultaneously*. For example, define:

$$\begin{cases} f:[0,2\pi]\to\mathbb{R}^2 \\ f(x)=(\cos(x),\sin(x)) \end{cases}$$

Then $f(2\pi)-f(0)=\mathbf{0}\in\mathbb{R}^2$, but $f_1'(x)=-\sin(x)$ and $f_2'(x)=\cos(x)$ are never simultaneously zero as x ranges over $[0,2\pi]$.

However a certain type of generalization of the mean value theorem to vector-valued functions is obtained as follows: Let f be a continuously differentiable real-valued function defined on an open interval I, and let x as well as $x+h$ be points of I. The mean value theorem in one variable tells us that there exists some t^* between 0 and 1 such that,

$$f(x+h)-f(x)=f'(x+t^*h)\cdot h.$$

On the other hand, we have, by the fundamental theorem of calculus followed by a change of variables,

$$f(x+h)-f(x)=\int_x^{x+h}f'(u)du=\left(\int_0^1 f'(x+th)dt\right)\cdot h.$$

Thus, the value $f'(x+t^*h)$ at the particular point t^* has been replaced by the mean value,

$$\int_0^1 f'(x+th)dt.$$

This last version can be generalized to vector valued functions:

Lemma: Let $U \subset \mathbf{R}^n$ be open, $f : U \to \mathbf{R}^m$ continuously differentiable, and $x \in U$, $h \in \mathbf{R}^n$ vectors such that the line segment $x + th$, $0 \leq t \leq 1$ remains in U. Then we have:

$$f(x+h) - f(x) = \left(\int_0^1 Df(x+th)dt \right) \cdot h,$$

where Df denotes the Jacobian matrix of f and the integral of a matrix is to be understood componentwise.

Proof: Let $f_1, \ldots, f_m$ denote the components of f and define-

$$\begin{cases} g_i : [0,1] \to \mathbf{R} \\ g_i(t) = f_i(x + th) \end{cases}$$

Then we have,

$$f_i(x+h) - f_i(x) = g_i(1) - g_i(0) = \int_0^1 g_{i'}(t)dt$$

$$= \int_0^1 \left(\sum_{j=1}^n \frac{\partial f_i}{\partial x_j}(x+th)h_j \right) dt = \sum_{j=1}^n \left(\int_0^1 \frac{\partial f_i}{\partial x_j}(x+th)dt \right) h_j.$$

The claim follows since Df is the matrix consisting of the components $\dfrac{\partial f_i}{\partial x_j}$.

Lemma: Let $v : [a,b] \to \mathbf{R}^m$ be a continuous function defined on the interval $[a,b] \subset \mathbf{R}$. Then we have,

$$\left\| \int_a^b v(t)\, dt \right\| \leq \int_a^b \| v(t) \|\, dt.$$

Proof: Let u in $\mathbf{R}^m$ denote the value of the integral:

$$u := \int_a^b v(t)dt.$$

Now we have (using the Cauchy–Schwarz inequality):

$$\| u \|^2 = \langle u, u \rangle = \left\langle \int_a^b v(t)dt, u \right\rangle = \int_a^b \langle v(t), u \rangle\, dt \leq \int_a^b \| v(t) \| \cdot \| u \|\, dt = \| u \| \int_a^b \| v(t) \|\, dt$$

Now cancelling the norm of u from both ends gives us the desired inequality.

Mean Value Inequality. If the norm of $Df(x+th)$ is bounded by some constant M for t in [0, 1], then,

$$\| f(x+h) - f(x) \| \leq M \| h \|.$$

Proof: From Lemma first and second it follows that,

$$\| f(x+h) - f(x) \| = \left\| \int_0^1 (Df(x+th) \cdot h) dt \right\| \leqslant \int_0^1 \| Df(x+th) \| \cdot \| h \| dt \leqslant M \| h \|.$$

Mean Value Theorems for Definite Integrals

First mean value theorem for definite integrals:

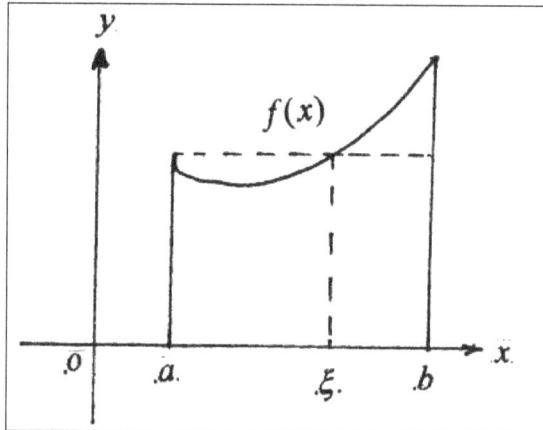

Geometrically: interpreting f(c) as the height of a rectangle and $b-a$ as the width,
this rectangle has the same area as the region below the curve from a to b.

Let $f : [a,b] \to \mathbf{R}$ be a continuous function. Then there exists c in (a, b) such that,

$$\int_a^b f(x) dx = f(c)(b-a).$$

Since the mean value of f on $[a, b]$ is defined as,

$$\frac{1}{b-a} \int_a^b f(x) dx,$$

we can interpret the conclusion as f achieves its mean value at some c in (a, b).

In general, if $f : [a,b] \to \mathbf{R}$ is continuous and g is an integrable function that does not change sign on $[a, b]$, then there exists c in (a, b) such that

$$\int_a^b f(x) g(x) dx = f(c) \int_a^b g(x) dx.$$

Proof of the first mean value theorem for definite integrals:

Suppose $f : [a,b] \to \mathbf{R}$ is continuous and g is a nonnegative integrable function on $[a, b]$. By the extreme value theorem, there exists m and M such that for each x in $[a, b]$, $m \leqslant f(x) \leqslant M$ and $f[a,b] = [m, M]$. Since g is nonnegative,

$$m \int_a^b g(x) dx \leqslant \int_a^b f(x) g(x) dx \leqslant M \int_a^b g(x) dx.$$

Now let,

$$I = \int_a^b g(x)dx.$$

If $I = 0$, we're done since,

$$0 \leqslant \int_a^b f(x)g(x)\,dx \leqslant 0$$

means,

$$\int_a^b f(x)g(x)\,dx = 0,$$

so for any c in (a, b),

$$\int_a^b f(x)g(x)\,dx = f(c)I = 0.$$

If $I \neq 0$, then,

$$m \leqslant \frac{1}{I}\int_a^b f(x)g(x)\,dx \leqslant M.$$

By the intermediate value theorem, f attains every value of the interval $[m, M]$, so for some c in $[a, b]$,

$$f(c) = \frac{1}{I}\int_a^b f(x)g(x)dx,$$

that is,

$$\int_a^b f(x)g(x)\,dx = f(c)\int_a^b g(x)\,dx.$$

Finally, if g is negative on $[a, b]$, then,

$$M\int_a^b g(x)\,dx \leqslant \int_a^b f(x)g(x)\,dx \leqslant m\int_a^b g(x)dx,$$

and we still get the same result as above.

Second mean value theorem for definite integrals:

There are various slightly different theorems called the second mean value theorem for definite integrals. A commonly found version is as follows.

If $G : [a,b] \to \mathbf{R}$ is a positive monotonically decreasing function and $\varphi : [a,b] \to \mathbf{R}$ is an integrable function, then there exists a number x in $(a, b]$ such that,

$$\int_a^b G(t)\varphi(t)\,dt = G(a^+)\int_a^x \varphi(t)\,dt.$$

Here $G(a^+)$ stands for $\lim\limits_{x \to a^+} G(x)$, the existence of which follows from the conditions. Note that it is essential that the interval $(a, b]$ contains b. A variant not having this requirement is:

If $G : [a,b] \to \mathbf{R}$ is a monotonic (not necessarily decreasing and positive) function and $\varphi : [a,b] \to \mathbf{R}$ is an integrable function, then there exists a number x in (a, b) such that,

$$\int_a^b G(t)\varphi(t)\ dt = G(a^+) \int_a^x \varphi(t)dt + G(b^-) \int_x^b \varphi(t)dt.$$

Mean Value Theorem for Integration Fails for Vector-valued Functions

If the function G returns a multi-dimensional vector, then the MVT for integration is not true, even if the domain of G is also multi-dimensional.

For example, consider the following 2-dimensional function defined on an n-dimensional cube:

$$\begin{cases} G : [0,2\pi]^n \to \mathbb{R}^2 \\ G(x_1,\cdots,x_n) = \left(\sin(x_1 + \cdots + x_n), \cos(x_1 + \cdots + x_n)\right) \end{cases}$$

Then, by symmetry it is easy to see that the mean value of G over its domain is $(0,0)$:

$$\int_{[0,2\pi]^n} G(x_1,\cdots,x_n)dx_1 \cdots dx_n = (0,0)$$

However, there is no point in which $G = (0,0)$, because $|G| = 1$ everywhere.

Probabilistic Analogue of the Mean Value Theorem

Let X and Y be non-negative random variables such that $E[X] < E[Y] < \infty$ and $X \leqslant_{st} Y$ (i.e. X is smaller than Y in the usual stochastic order). Then there exists an absolutely continuous non-negative random variable Z having probability density function,

$$f_Z(x) = \frac{\Pr(Y > x) - \Pr(X > x)}{E[Y] - E[X]}, \qquad x \geqslant 0.$$

Let g be a measurable and differentiable function such that $E\big[g(X)\big], E\big[g(Y)\big] < \infty$, and let its derivative g' be measurable and Riemann-integrable on the interval $[x, y]$ for all $y \geq x \geq 0$. Then, $E\big[g'(Z)\big]$ is finite and

$$E[g(Y)] - E[g(X)] = E[g'(Z)][E(Y) - E(X)].$$

Generalization in Complex Analysis

As noted above, the theorem does not hold for differentiable complex-valued functions. Instead, a generalization of the theorem is stated such:

Let $f : \Omega \to \mathbf{C}$ be a holomorphic function on the open convex set Ω, and let a and b be distinct points in Ω. Then there exist points u, v on L_{ab} (the line segment from a to b) such that,

$$Re(f'(u)) = Re\left(\frac{f(b) - f(a)}{b - a}\right),$$

$$Im(f'(v)) = Im\left(\frac{f(b) - f(a)}{b - a}\right).$$

Where Re() is the Real part and Im() is the Imaginary part of a complex-valued function.

FERMAT'S THEOREM

In mathematics, Fermat's theorem (also known as interior extremum theorem) is a method to find local maxima and minima of differentiable functions on open sets by showing that every local extremum of the function is a stationary point (the function derivative is zero at that point). Fermat's theorem is a theorem in real analysis, named after Pierre de Fermat.

By using Fermat's theorem, the potential extrema of a function f, with derivative f', are found by solving an equation in f'. Fermat's theorem gives only a necessary condition for extreme function values, as some stationary points are inflection points (not a maximum or minimum). The function's second derivative, if it exists, can determine if any stationary point is a maximum, minimum, or inflection point.

One way to state Fermat's theorem is that, if a function has a local extremum at some point and is differentiable there, then the function's derivative at that point must be zero. In precise mathematical language:

Let $f : (a, b) \to \mathbb{R}$ be a function and suppose that $x_0 \in (a, b)$ is a point where f has a local extremum. If f is differentiable at x_0, then $f'(x_0) = 0$.

Another way to understand the theorem is via the contrapositive statement: if the derivative of a function at any point is not zero, then there is not a local extremum at that point. Formally:

If f is differentiable at $x_0 \in (a, b)$, and $f'(x_0) \neq 0$, then x_0 is not a local extremum of f.

Corollary

The global extrema of a function f on a domain A occur only at boundaries, non-differentiable points, and stationary points. If x_0 is a global extremum of f, then one of the following is true:

- Boundary: x_0 is in the boundary of A.

- Non-differentiable: f is not differentiable at x_0.

- Stationary point: x_0 is a stationary point of f.

Extension

In higher dimensions, exactly the same statement holds; however, the proof is slightly more complicated. The complication is that in 1 dimension, one can either move left or right from a point, while in higher dimensions, one can move in many directions. Thus, if the derivative does not vanish, one must argue that there is *some* direction in which the function increases – and thus in the opposite direction the function decreases. This is the only change to the proof or the analysis.

The statement can also be extended to differentiable manifolds. If $f : M \to \mathbb{R}$ is a differentiable function on a manifold M, then its local extrema must be critical points of f, in particular points where the exterior derivative df is zero.

Applications

Fermat's theorem is central to the calculus method of determining maxima and minima: in one dimension, one can find extrema by simply computing the stationary points (by computing the zeros of the derivative), the non-differentiable points, and the boundary points, and then investigating this set to determine the extrema.

One can do this either by evaluating the function at each point and taking the maximum, or by analyzing the derivatives further, using the first derivative test, the second derivative test, or the higher-order derivative test.

Intuitive Argument

Intuitively, a differentiable function is approximated by its derivative – a differentiable function behaves infinitesimally like a linear function $a + bx$, or more precisely, $f(x_0) + f'(x_0) \cdot (x - x_0)$. Thus, from the perspective that "if f is differentiable and has non-vanishing derivative at x_0, then it does not attain an extremum at x_0," the intuition is that if the derivative at x_0 is positive, the function is *increasing* near x_0, while if the derivative is negative, the function is *decreasing* near x_0. In both cases, it cannot attain a maximum or minimum, because its value is changing. It can only attain a maximum or minimum if it "stops" – if the derivative vanishes (or if it is not differentiable, or if one runs into the boundary and cannot continue). However, making "behaves like a linear function" precise requires careful analytic proof.

More precisely, the intuition can be stated as: if the derivative is positive, there is *some point* to the right of x_0 where f is greater, and *some point* to the left of x_0 where f is less, and thus f attains neither a maximum nor a minimum at x_0 Conversely, if the derivative is negative, there is a point to the right which is lesser, and a point to the left which is greater. Stated this way, the proof is just translating this into equations and verifying "how much greater or less".

The intuition is based on the behavior of polynomial functions. Assume that function f has a maximum at x_0, the reasoning being similar for a function minimum. If $x_0 \in (a,b)$ is a local maximum then, roughly, there is a (possibly small) neighborhood of x_0 such as the function "is increasing before" and "decreasing after" x_0. As the derivative is positive for an increasing function and negative for a decreasing function, f' is positive before and negative after x_0. f' doesn't skip values (by Darboux's theorem), so it has to be zero at some point between the positive and negative values. The only point in the neighbourhood where it is possible to have $f'(x) = 0$ is x_0.

The theorem (and its proof below) is more general than the intuition in that it doesn't require the function to be differentiable over a neighbourhood around x_0. It is sufficient for the function to be differentiable only in the extreme point.

Proof: Non-vanishing derivatives implies not extremum.

Suppose that f is differentiable at $x_0 \in (a,b)$, with derivative K, and assume without loss of generality that $K > 0$, so the tangent line at x_0 has positive slope (is increasing). Then there is a neighborhood of x_0 on which the secant lines through x_0 all have positive slope, and thus to the right of x_0 f is greater, and to the left of x_0, f is lesser.

The schematic of the proof is:

- An infinitesimal statement about derivative (tangent line) *at* x_0 implies.

- A local statement about difference quotients (secant lines) *near* x_0, which implies.

- A local statement about the *value* of f near x_0.

Formally, by the definition of derivative, $f'(x_0) = K$ means that:

$$\lim_{\varepsilon \to 0} \frac{f(x_0 + \varepsilon) - f(x_0)}{\varepsilon} = K.$$

In particular, for sufficiently small ε (less than some ε_0), the fraction must be at least $K/2$, by the definition of limit. Thus on the interval $(x_0 - \varepsilon_0, x_0 + \varepsilon_0)$ one has:

$$\frac{f(x_0 + \varepsilon) - f(x_0)}{\varepsilon} > K/2;$$

one has replaced the *equality* in the limit (an infinitesimal statement) with an *inequality* on a neighborhood (a local statement). Thus, rearranging the equation, if $\varepsilon > 0$ then:

$$f(x_0 + \varepsilon) > f(x_0) + (K/2)\varepsilon > f(x_0),$$

so on the interval to the right, f is greater than $f(x_0)$, and if $\varepsilon < 0$, then:

$$f(x_0 + \varepsilon) < f(x_0) + (K/2)\varepsilon < f(x_0),$$

so on the interval to the left, f is less than $f(x_0)$.

Thus x_0 is not a local or global maximum or minimum of f.

Proof: Extremum implies derivative vanishes.

Alternatively, one can start by assuming that x_0 is a local maximum, and then prove that the derivative is 0.

Suppose that x_0 is a local maximum (a similar proof applies if x_0 is a local minimum). Then there $\exists \delta > 0$ such that $(x_0 - \delta, x_0 + \delta) \subset (a,b)$ and such that we have $f(x_0) \geq f(x) \forall x$ with $|x - x_0| < \delta$. Hence for any $h \in (0, \delta)$ we notice that it holds:

$$\frac{f(x_0 + h) - f(x_0)}{h} \leq 0.$$

Since the limit of this ratio as h gets close to 0 from above exists and is equal to $f'(x_0)$ we conclude that $f'(x_0) \leq 0$. On the other hand, for $h \in (-\delta, 0)$ we notice that:

$$\frac{f(x_0 + h) - f(x_0)}{h} \geq 0$$

but again the limit as h gets close to 0 from below exists and is equal to $f'(x_0)$ so we also have $f'(x_0) \geq 0$.

Hence we conclude that $f'(x_0) = 0$.

Cautions

A subtle misconception that is often held in the context of Fermat's theorem is to assume that it makes a stronger statement about local behavior than it does. Notably, Fermat's theorem does *not* say that functions (monotonically) "increase up to" or "decrease down from" a local maximum. This is very similar to the misconception that a limit means "monotonically getting closer to a point". For "well-behaved functions" (which here mean continuously differentiable), some intuitions hold, but in general functions may be ill-behaved, as illustrated below. The moral is that derivatives determine *infinitesimal* behavior, and that *continuous* derivatives determine *local* behavior.

Continuously Differentiable Functions

If *f* is continuously differentiable $\left(C^1 \right)$ on an open neighborhood of the point x_0, then $f'(x_0) > 0$ does mean that *f* is increasing on a neighborhood of x_0, as follows.

If $f'(x_0) = K > 0$ and $f \in C^1$, then by continuity of the derivative, there is some $\varepsilon_0 > 0$ such that $f'(x_0) > K / 2 \ \forall x \in \left(x_0 - \varepsilon_0, x_0 + \varepsilon_0 \right)$. Then *f* is increasing on this interval, by the mean value theorem: the slope of any secant line is at least $K / 2$, as it equals the slope of some tangent line.

However, in the general statement of Fermat's theorem, where one is only given that the derivative *at* x_0 is positive, one can only conclude that secant lines *through* x_0 will have positive slope, for secant lines between x_0 and near enough points.

Conversely, if the derivative of *f* at a point is zero (x_0 is a stationary point), one cannot in general conclude anything about the local behavior of *f* – it may increase to one side and decrease to the other (as in x^3), increase to both sides (as in x^4), decrease to both sides (as in $-x^4$), or behave in more complicated ways, such as oscillating as in $x^2(\sin(1/x))$.

One can analyze the infinitesimal behavior via the second derivative test and higher-order derivative test, if the function is differentiable enough, and if the first non-vanishing derivative at x_0 is a continuous function, one can then conclude local behavior (i.e., if $f^{(k)}(x_0) \neq 0$ is the first non-vanishing derivative, and $f^{(k)}$ is continuous, so $f \in C^k$), then one can treat f as locally close to a polynomial of degree k, since it behaves approximately as $f^{(k)}(x_0)(x - x_0)^k$, but if the kth derivative is not continuous, one cannot draw such conclusions, and it may behave rather differently.

Pathological Functions

Consider the function $\sin(1/x)$ – it oscillates increasingly rapidly between -1 and 1 as x approaches 0. Consider then $f(x) = (1 + \sin(1/x))x^2$ – this oscillates increasingly rapidly between 0 and $2x^2$ as x approaches 0. If one extends this function by $f(0) := 0$, then the function is continuous and everywhere differentiable (it is differentiable at 0 with derivative 0), but has rather unexpected behavior near 0: in any neighborhood of 0 it attains 0 infinitely many times, but also equals $2x^2$ (a positive number) infinitely often.

Continuing in this vein, $f(x) = (2 + \sin(1/x))x^2$ oscillates between x^2 and $3x^2$ and $x = 0$ is a local and global minimum, but on no neighborhood of 0 is it decreasing down to or increasing up from 0 – it oscillates wildly near 0.

This pathology can be understood because, while the function is everywhere differentiable, it is not *continuously* differentiable: the limit of $f'(x)$ as $x \to 0$ does not exist, so the derivative is not continuous at 0. This reflects the oscillation between increasing and decreasing values as it approaches 0.

TAYLOR'S SERIES

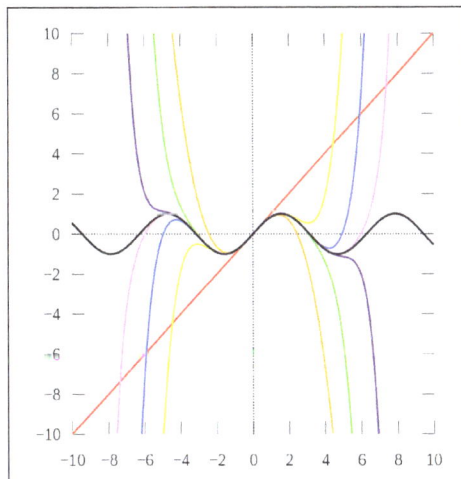

As the degree of the Taylor polynomial rises, it approaches the correct function. This image shows $\sin x$ and its Taylor approximations, polynomials of degree 1, 3, 5, 7, 9, 11 and 13.

In mathematics, a Taylor series is a representation of a function as an infinite sum of terms that are calculated from the values of the function's derivatives at a single point.

In the West, the subject was formulated by the Scottish mathematician James Gregory and formally introduced by the English mathematician Brook Taylor in 1715. If the Taylor series is centered at zero, then that series is also called a Maclaurin series, after the Scottish mathematician Colin Maclaurin, who made extensive use of this special case of Taylor series in the 18th century.

A function can be approximated by using a finite number of terms of its Taylor series. Taylor's theorem gives quantitative estimates on the error introduced by the use of such an approximation. The polynomial formed by taking some initial terms of the Taylor series is called a Taylor polynomial. The Taylor series of a function is the limit of that function's Taylor polynomials as the degree increases, provided that the limit exists. A function may not be equal to its Taylor series, even if its Taylor series converges at every point. A function that is equal to its Taylor series in an open interval (or a disc in the complex plane) is known as an analytic function in that interval.

The Taylor series of a real or complex-valued function $f(x)$ that is infinitely differentiable at a real or complex number a is the power series:

$$f(a) + \frac{f'(a)}{1!}(x-a) + \frac{f''(a)}{2!}(x-a)^2 + \frac{f'''(a)}{3!}(x-a)^3 + \cdots,$$

where $n!$ denotes the factorial of n and $f^{(n)}(a)$ denotes the nth derivative of f evaluated at the point a. In the more compact sigma notation, this can be written as:

$$\sum_{n=0}^{\infty} \frac{f^{(n)}(a)}{n!}(x-a)^n.$$

The derivative of order zero of f is defined to be f itself and $(x-a)^0$ and $0!$ are both defined to be 1. When $a = 0$, the series is also called a Maclaurin series.

The Taylor series for any polynomial is the polynomial itself.

The Maclaurin series for $\dfrac{1}{1-x}$ is the geometric series:

$$1 + x + x^2 + x^3 + \cdots$$

so the Taylor series for $\dfrac{1}{x}$ at $a = 1$ is:

$$1 - (x-1) + (x-1)^2 - (x-1)^3 + \cdots.$$

By integrating the above Maclaurin series, we find the Maclaurin series for $\ln(1-x)$, where log denotes the natural logarithm:

$$-x - \tfrac{1}{2}x^2 - \tfrac{1}{3}x^3 - \tfrac{1}{4}x^4 - \cdots$$

and the corresponding Taylor series for $\ln x$ at $a = 1$ is,

$$(x-1) - \tfrac{1}{2}(x-1)^2 + \tfrac{1}{3}(x-1)^3 - \tfrac{1}{4}(x-1)^4 + \cdots,$$

and more generally, the corresponding Taylor series for $\ln x$ at some $a = x_0$ is:

$$\ln x_0 + \frac{1}{x_0}(x - x_0) - \frac{1}{x_0^2}\frac{(x - x_0)^2}{2} + \cdots.$$

The Taylor series for the exponential function e^x at $a = 0$ is,

$$\sum_{n=0}^{\infty} \frac{x^n}{n!} = \frac{x^0}{0!} + \frac{x^1}{1!} + \frac{x^2}{2!} + \frac{x^3}{3!} + \frac{x^4}{4!} + \frac{x^5}{5!} + \cdots$$

$$= 1 + x + \frac{x^2}{2} + \frac{x^3}{6} + \frac{x^4}{24} + \frac{x^5}{120} + \cdots.$$

The above expansion holds because the derivative of e^x with respect to x is also e^x and e^0 equals 1. This leaves the terms $(x - 0)^n$ in the numerator and $n!$ in the denominator for each term in the infinite sum.

Analytic Functions

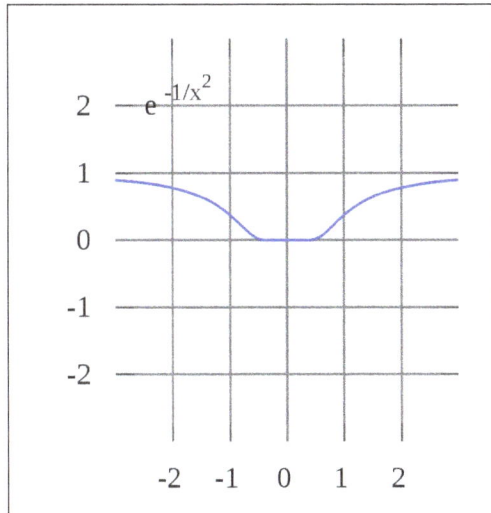

The function $e^{(-1/x^2)}$ is not analytic at $x - 0$: the Taylor series is identically 0, although the function is not.

If $f(x)$ is given by a convergent power series in an open disc (or interval in the real line) centred at b in the complex plane, it is said to be analytic in this disc. Thus for x in this disc, f is given by a convergent power series:

$$f(x) = \sum_{n=0}^{\infty} a_n (x - b)^n.$$

Differentiating by x the above formula n times, then setting $x = b$ gives:

$$\frac{f^{(n)}(b)}{n!} = a_n$$

and so the power series expansion agrees with the Taylor series. Thus a function is analytic in an open disc centred at b if and only if its Taylor series converges to the value of the function at each point of the disc.

If $f(x)$ is equal to its Taylor series for all x in the complex plane, it is called entire. The polynomials, exponential function e^x, and the trigonometric functions sine and cosine, are examples of entire functions. Examples of functions that are not entire include the square root, the logarithm, the trigonometric function tangent, and its inverse, arctan. For these functions the Taylor series do not converge if x is far from b. That is, the Taylor series diverges at x if the distance between x and b is larger than the radius of convergence. The Taylor series can be used to calculate the value of an entire function at every point, if the value of the function, and of all of its derivatives, are known at a single point.

Uses of the Taylor series for analytic functions include:

1. The partial sums (the Taylor polynomials) of the series can be used as approximations of the function. These approximations are good if sufficiently many terms are included.

2. Differentiation and integration of power series can be performed term by term and is hence particularly easy.

3. An analytic function is uniquely extended to a holomorphic function on an open disk in the complex plane. This makes the machinery of complex analysis available.

4. The (truncated) series can be used to compute function values numerically, (often by recasting the polynomial into the Chebyshev form and evaluating it with the Clenshaw algorithm).

5. Algebraic operations can be done readily on the power series representation; for instance, Euler's formula follows from Taylor series expansions for trigonometric and exponential functions. This result is of fundamental importance in such fields as harmonic analysis.

6. Approximations using the first few terms of a Taylor series can make otherwise unsolvable problems possible for a restricted domain; this approach is often used in physics.

Approximation Error and Convergence

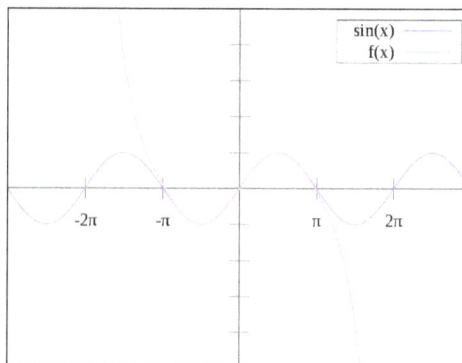

The sine function (blue) is closely approximated by its Taylor polynomial of degree 7 (pink) for a full period centered at the origin.

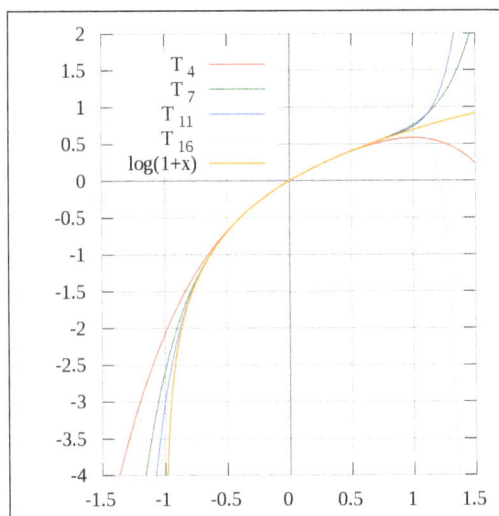

The Taylor polynomials for $\log(1 + x)$ only provide accurate approximations in the range $-1 < x \leq 1$. For $x > 1$, Taylor polynomials of higher degree provide worse approximations.

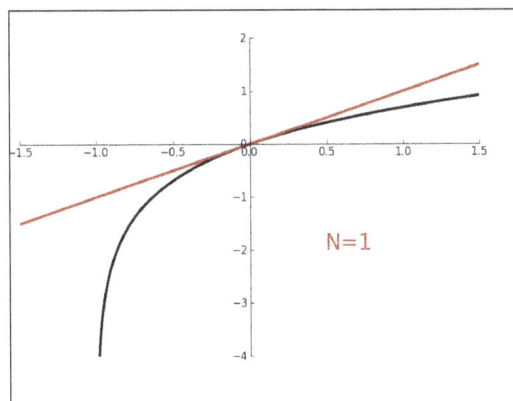

The Taylor approximations for $\log(1 + x)$ (black).
For $x > 1$, the approximations diverge.

Figure above is an accurate approximation of $\sin x$ around the point $x = 0$. The pink curve is a polynomial of degree seven:

$$\sin(x) \approx x - \frac{x^3}{3!} + \frac{x^5}{5!} - \frac{x^7}{7!}.$$

The error in this approximation is no more than $\dfrac{|x|^9}{9!}$. In particular, for $-1 < x < 1$, the error is less than 0.000003.

In contrast, also shown is a picture of the natural logarithm function $\log(1 + x)$ and some of its Taylor polynomials around $a = 0$. These approximations converge to the function only in the region $-1 < x \leq 1$; outside of this region the higher-degree Taylor polynomials are *worse* approximations for the function. This is similar to Runge's phenomenon.

The *error* incurred in approximating a function by its nth-degree Taylor polynomial is called the *remainder* or *residual* and is denoted by the function $R_n(x)$. Taylor's theorem can be used to obtain a bound on the size of the remainder.

In general, Taylor series need not be convergent at all. And in fact the set of functions with a convergent Taylor series is a meager set in the Fréchet space of smooth functions. And even if the Taylor series of a function f does converge, its limit need not in general be equal to the value of the function $f(x)$. For example, the function:

$$f(x) = \begin{cases} e^{-\frac{1}{x^2}} & \text{if } x \neq 0 \\ 0 & \text{if } x = 0 \end{cases}$$

is infinitely differentiable at $x = 0$, and has all derivatives zero there. Consequently, the Taylor series of $f(x)$ about $x = 0$ is identically zero. However, $f(x)$ is not the zero function, so does not equal its Taylor series around the origin. Thus, $f(x)$ is an example of a non-analytic smooth function.

In real analysis, this example shows that there are infinitely differentiable functions $f(x)$ whose Taylor series are *not* equal to $f(x)$ even if they converge. By contrast, the holomorphic functions studied in complex analysis always possess a convergent Taylor series, and even the Taylor series of meromorphic functions, which might have singularities, never converge to a value different from the function itself. The complex function $e^{-1/z2}$, however, does not approach 0 when z approaches 0 along the imaginary axis, so it is not continuous in the complex plane and its Taylor series is undefined at 0.

More generally, every sequence of real or complex numbers can appear as coefficients in the Taylor series of an infinitely differentiable function defined on the real line, a consequence of Borel's lemma. As a result, the radius of convergence of a Taylor series can be zero. There are even infinitely differentiable functions defined on the real line whose Taylor series have a radius of convergence 0 everywhere.

A function cannot be written as a Taylor series centred at a singularity; in these cases, one can often still achieve a series expansion if one allows also negative powers of the variable x;. For example, $f(x) = e^{-1/x^2}$ can be written as a Laurent series.

Generalization

There is, however, a generalization of the Taylor series that does converge to the value of the function itself for any bounded continuous function on $(0, \infty)$, using the calculus of finite differences. Specifically, one has the following theorem, due to Einar Hille, that for any $t > 0$,

$$\lim_{h \to 0^+} \sum_{n=0}^{\infty} \frac{t^n}{n!} \frac{\Delta_h^n f(a)}{h^n} = f(a+t).$$

Here Δ_h^n is the nth finite difference operator with step size h. The series is precisely the Taylor series, except that divided differences appear in place of differentiation: the series is formally similar to the Newton series. When the function f is analytic at a, the terms in the series converge to the terms of the Taylor series, and in this sense generalizes the usual Taylor series.

In general, for any infinite sequence a_i, the following power series identity holds:

$$\sum_{n=0}^{\infty} \frac{u^n}{n!} \Delta^n a_i = e^{-u} \sum_{j=0}^{\infty} \frac{u^j}{j!} a_{i+j}.$$

So in particular,

$$f(a+t) = \lim_{h \to 0^+} e^{-\frac{t}{h}} \sum_{j=0}^{\infty} f(a+jh) \frac{\left(\dfrac{t}{h}\right)^j}{j!}.$$

The series on the right is the expectation value of $f(a+X)$, where X is a Poisson-distributed random variable that takes the value jh with probability $e^{-t/h} \cdot \dfrac{(t/h)^j}{j!}$. Hence,

$$f(a+t) = \lim_{h \to 0^+} \int_{-\infty}^{\infty} f(a+x) dP_{\frac{t}{h},h}(x).$$

The law of large numbers implies that the identity holds.

List of Maclaurin Series of some Common Functions

Several important Maclaurin series expansions follow. All these expansions are valid for complex arguments x.

Exponential Function

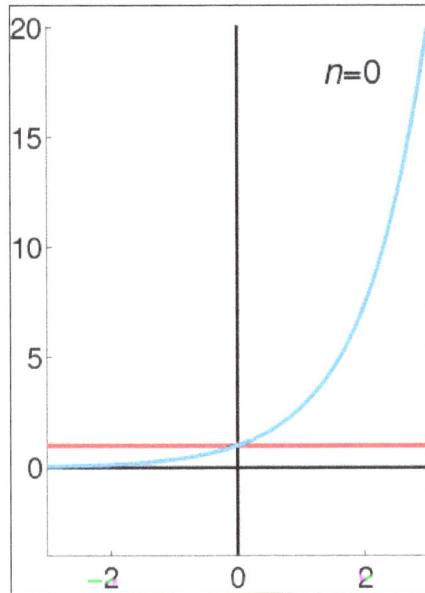

The exponential function e^x (in blue), and the sum of the first $n+1$ terms of its Taylor series at 0 (in red).

The exponential function e^x (with base e) has Maclaurin series:

$$e^x = \sum_{n=0}^{\infty} \frac{x^n}{n!} = 1 + x + \frac{x^2}{2!} + \frac{x^3}{3!} + \cdots.$$

It converges for all x.

Natural Logarithm

The natural logarithm (with base e) has Maclaurin series:

$$\log(1-x) = -\sum_{n=1}^{\infty} \frac{x^n}{n} = -x - \frac{x^2}{2} - \frac{x^3}{3} - \cdots,$$

$$\log(1+x) = \sum_{n=1}^{\infty} (-1)^{n+1} \frac{x^n}{n} = x - \frac{x^2}{2} + \frac{x^3}{3} - \cdots.$$

They converge for $|x| < 1$. Also log(1-x) converges for $x = -1$ and log(1+x) converges for $x = 1$.

Geometric Series

The geometric series and its derivatives have Maclaurin series

$$\frac{1}{1-x} = \sum_{n=0}^{\infty} x^n$$

$$\frac{1}{(1-x)^2} = \sum_{n=1}^{\infty} n x^{n-1}$$

$$\frac{1}{(1-x)^3} = \sum_{n=2}^{\infty} \frac{(n-1)n}{2} x^{n-2}.$$

All are convergent for $|x| < 1$. These are special cases of the binomial series.

Binomial Series

The binomial series is the power series

$$(1+x)^{\alpha} = \sum_{n=0}^{\infty} \binom{\alpha}{n} x^n$$

whose coefficients are the generalized binomial coefficients

$$\binom{\alpha}{n} = \prod_{k=1}^{n} \frac{\alpha - k + 1}{k} = \frac{\alpha(\alpha-1)\cdots(\alpha-n+1)}{n!}.$$

(If n = 0, this product is an empty product and has value 1.) It converges for $|x| < 1$ for any real or complex number α.

When $\alpha = -1$, this is essentially the infinite geometric series. The special cases $\alpha = \frac{1}{2}$ and $\alpha = \frac{-1}{2}$ give the square root function and its inverse:

$$(1+x)^{\frac{1}{2}} = 1 + \tfrac{1}{2}x - \tfrac{1}{8}x^2 + \tfrac{1}{16}x^3 - \tfrac{5}{128}x^4 + \tfrac{7}{256}x^5 - \cdots,$$

$$(1+x)^{-\frac{1}{2}} = 1 - \tfrac{1}{2}x + \tfrac{3}{8}x^2 - \tfrac{5}{16}x^3 + \tfrac{35}{128}x^4 - \tfrac{63}{256}x^5 + \cdots.$$

When only the linear term is retained, this simplifies to the binomial approximation.

Trigonometric Functions

The usual trigonometric functions and their inverses have the following Maclaurin series:

$$\sin x = \sum_{n=0}^{\infty} \frac{(-1)^n}{(2n+1)!} x^{2n+1} \qquad = x - \frac{x^3}{3!} + \frac{x^5}{5!} - \cdots \qquad \text{for all } x$$

$$\cos x = \sum_{n=0}^{\infty} \frac{(-1)^n}{(2n)!} x^{2n} \qquad = 1 - \frac{x^2}{2!} + \frac{x^4}{4!} - \cdots \qquad \text{for all } x$$

$$\tan x = \sum_{n=1}^{\infty} \frac{B_{2n}(-4)^n \left(1 - 4^n\right)}{(2n)!} x^{2n-1} \qquad = x + \frac{x^3}{3} + \frac{2x^5}{15} + \cdots \qquad \text{for } |x| < \frac{\pi}{2}$$

$$\sec x = \sum_{n=0}^{\infty} \frac{(-1)^n E_{2n}}{(2n)!} x^{2n} \qquad = 1 + \frac{x^2}{2} + \frac{5x^4}{24} + \cdots \qquad \text{for } |x| < \frac{\pi}{2}$$

$$\arcsin x = \sum_{n=0}^{\infty} \frac{(2n)!}{4^n (n!)^2 (2n+1)} x^{2n+1} \qquad = x + \frac{x^3}{6} + \frac{3x^5}{40} + \cdots \qquad \text{for } |x| \le 1$$

$$\arccos x = \frac{\pi}{2} - \arcsin x$$
$$= \frac{\pi}{2} - \sum_{n=0}^{\infty} \frac{(2n)!}{4^n (n!)^2 (2n+1)} x^{2n+1} \qquad = \frac{\pi}{2} - x - \frac{x^3}{6} - \frac{3x^5}{40} - \cdots \qquad \text{for } |x| \le 1$$

$$\arctan x = \sum_{n=0}^{\infty} \frac{(-1)^n}{2n+1} x^{2n+1} \qquad = x - \frac{x^3}{3} + \frac{x^5}{5} - \cdots \qquad \text{for } |x| \le 1, x \ne \pm i$$

All angles are expressed in radians. The numbers B_k appearing in the expansions of $\tan x$ are the Bernoulli numbers. The E_k in the expansion of $\sec x$ are Euler numbers.

Hyperbolic Functions

The hyperbolic functions have Maclaurin series closely related to the series for the corresponding trigonometric functions:

$$\sinh x = \sum_{n=0}^{\infty} \frac{x^{2n+1}}{(2n+1)!} \qquad = x + \frac{x^3}{3!} + \frac{x^5}{5!} + \cdots \qquad \text{for all } x$$

$$\cosh x = \sum_{n=0}^{\infty} \frac{x^{2n}}{(2n)!} \qquad = 1 + \frac{x^2}{2!} + \frac{x^4}{4!} + \cdots \qquad \text{for all } x$$

$$\tanh x = \sum_{n=1}^{\infty} \frac{B_{2n} 4^n \left(4^n - 1\right)}{(2n)!} x^{2n-1} \qquad = x - \frac{x^3}{3} + \frac{2x^5}{15} - \frac{17x^7}{315} + \cdots \qquad \text{for } |x| < \frac{\pi}{2}$$

$$\operatorname{arsinh} x = \sum_{n=0}^{\infty} \frac{(-1)^n (2n)!}{4^n (n!)^2 (2n+1)} x^{2n+1} \qquad \text{for } |x| \le 1$$

$$\operatorname{artanh} x \sum_{n=0}^{\infty} \frac{x^{2n+1}}{(2n+1)} \qquad \text{for } |x| \le 1, x \ne \pm 1$$

The numbers B_k appearing in the series for $\tanh x$ are the Bernoulli numbers.

Calculation of Taylor series

Several methods exist for the calculation of Taylor series of a large number of functions. One can attempt to use the definition of the Taylor series, though this often requires generalizing the form of the coefficients according to a readily apparent pattern. Alternatively, one can use manipulations such as substitution, multiplication or division, addition or subtraction of standard Taylor series to construct the Taylor series of a function, by virtue of Taylor series being power series. In some cases, one can also derive the Taylor series by repeatedly applying integration by parts. Particularly convenient is the use of computer algebra systems to calculate Taylor series.

First example:

In order to compute the 7th degree Maclaurin polynomial for the function,

$$f(x) = \ln(\cos x), \quad x \in \left(-\frac{\pi}{2}, \frac{\pi}{2}\right),$$

one may first rewrite the function as,

$$f(x) = \ln\left(1 + (\cos x - 1)\right).$$

The Taylor series for the natural logarithm is (using the big O notation),

$$\ln(1+x) = x - \frac{x^2}{2} + \frac{x^3}{3} + O\left(x^4\right)$$

and for the cosine function,

$$\cos x - 1 = -\frac{x^2}{2} + \frac{x^4}{24} - \frac{x^6}{720} + O\left(x^8\right)$$

The latter series expansion has a zero constant term, which enables us to substitute the second series into the first one and to easily omit terms of higher order than the 7th degree by using the big O notation:

$$
\begin{aligned}
f(x) &= \ln\left(1 + (\cos x - 1)\right) \\
&= (\cos x - 1) - \tfrac{1}{2}(\cos x - 1)^2 + \tfrac{1}{3}(\cos x - 1)^3 + O\left((\cos x - 1)^4\right) \\
&= \left(-\frac{x^2}{2} + \frac{x^4}{24} - \frac{x^6}{720} + O\left(x^8\right)\right) - \frac{1}{2}\left(-\frac{x^2}{2} + \frac{x^4}{24} + O\left(x^6\right)\right)^2 + \frac{1}{3}\left(-\frac{x^2}{2} + O\left(x^4\right)\right)^3 + O\left(x^8\right) \\
&= -\frac{x^2}{2} + \frac{x^4}{24} - \frac{x^6}{720} - \frac{x^4}{8} + \frac{x^6}{48} - \frac{x^6}{24} + O\left(x^8\right) \\
&= -\frac{x^2}{2} - \frac{x^4}{12} - \frac{x^6}{45} + O\left(x^8\right).
\end{aligned}
$$

Since the cosine is an even function, the coefficients for all the odd powers $x, x^3, x^5, x^7, \ldots$ have to be zero.

Second example:

Suppose we want the Taylor series at 0 of the function,

$$g(x) = \frac{e^x}{\cos x}.$$

We have for the exponential function,

$$e^x = \sum_{n=0}^{\infty} \frac{x^n}{n!} = 1 + x + \frac{x^2}{2!} + \frac{x^3}{3!} + \frac{x^4}{4!} + \cdots$$

and, as in the first example,

$$\cos x = 1 - \frac{x^2}{2!} + \frac{x^4}{4!} - \cdots$$

Assume the power series is,

$$\frac{e^x}{\cos x} = c_0 + c_1 x + c_2 x^2 + c_3 x^3 + \cdots$$

Then multiplication with the denominator and substitution of the series of the cosine yields,

$$e^x = \left(c_0 + c_1 x + c_2 x^2 + c_3 x^3 + \cdots \right) \cos x$$

$$= \left(c_0 + c_1 x + c_2 x^2 + c_3 x^3 + c_4 x^4 + \cdots \right)\left(1 - \frac{x^2}{2!} + \frac{x^4}{4!} - \cdots \right)$$

$$= c_0 - \frac{c_0}{2} x^2 + \frac{c_0}{4!} x^4 + c_1 x - \frac{c_1}{2} x^3 + \frac{c_1}{4!} x^5 + c_2 x^2 - \frac{c_2}{2} x^4 + \frac{c_2}{4!} x^6 + c_3 x^3 - \frac{c_3}{2} x^5 + \frac{c_3}{4!} x^7 + c_4 x^4 + \cdots$$

Collecting the terms up to fourth order yields,

$$e^x = c_0 + c_1 x + \left(c_2 - \frac{c_0}{2} \right) x^2 + \left(c_3 - \frac{c_1}{2} \right) x^3 + \left(c_4 - \frac{c_2}{2} + \frac{c_0}{4!} \right) x^4 + \cdots$$

The values of c_i can be found by comparison of coefficients with the top expression for e^x, yielding:

$$\frac{e^x}{\cos x} = 1 + x + x^2 + \frac{2x^3}{3} + \frac{x^4}{2} + \cdots$$

Third example:

Here we employ a method called "indirect expansion" to expand the given function. This method uses the known Taylor expansion of the exponential function. In order to expand $(1 + x)e^x$ as a Taylor series in x, we use the known Taylor series of function e^x:

$$e^x = \sum_{n=0}^{\infty} \frac{x^n}{n!} = 1 + x + \frac{x^2}{2!} + \frac{x^3}{3!} + \frac{x^4}{4!} + \cdots$$

Thus,

$$(1+x)e^x = e^x + xe^x = \sum_{n=0}^{\infty} \frac{x^n}{n!} + \sum_{n=0}^{\infty} \frac{x^{n+1}}{n!} = 1 + \sum_{n=1}^{\infty} \frac{x^n}{n!} + \sum_{n=0}^{\infty} \frac{x^{n+1}}{n!}$$

$$= 1 + \sum_{n=1}^{\infty} \frac{x^n}{n!} + \sum_{n=1}^{\infty} \frac{x^n}{(n-1)!} = 1 + \sum_{n=1}^{\infty} \left(\frac{1}{n!} + \frac{1}{(n-1)!} \right) x^n$$

$$= 1 + \sum_{n=1}^{\infty} \frac{n+1}{n!} x^n$$

$$= \sum_{n=0}^{\infty} \frac{n+1}{n!} x^n.$$

Taylor Series as Definitions

Classically, algebraic functions are defined by an algebraic equation, and transcendental functions are defined by some property that holds for them, such as a differential equation. For example, the exponential function is the function which is equal to its own derivative everywhere, and assumes the value 1 at the origin. However, one may equally well define an analytic function by its Taylor series.

Taylor series are used to define functions and "operators" in diverse areas of mathematics. In particular, this is true in areas where the classical definitions of functions break down. For example, using Taylor series, one may extend analytic functions to sets of matrices and operators, such as the matrix exponential or matrix logarithm.

In other areas, such as formal analysis, it is more convenient to work directly with the power series themselves. Thus one may define a solution of a differential equation *as* a power series which, one hopes to prove, is the Taylor series of the desired solution.

Taylor Series in Several Variables

The Taylor series may also be generalized to functions of more than one variable with:

$$T(x_1,\ldots,x_d) = \sum_{n_1=0}^{\infty} \cdots \sum_{n_d=0}^{\infty} \frac{(x_1-a_1)^{n_1} \cdots (x_d-a_d)^{n_d}}{n_1! \cdots n_d!} \left(\frac{\partial^{n_1+\cdots+n_d} f}{\partial x_1^{n_1} \cdots \partial x_d^{n_d}} \right) (a_1,\ldots,a_d)$$

$$= f(a_1,\ldots,a_d) + \sum_{j=1}^{d} \frac{\partial f(a_1,\ldots,a_d)}{\partial x_j}(x_j-a_j) + \frac{1}{2!} \sum_{j=1}^{d} \sum_{k=1}^{d} \frac{\partial^2 f(a_1,\ldots,a_d)}{\partial x_j \partial x_k}(x_j-a_j)(x_k-a_k) +$$

$$+ \frac{1}{3!} \sum_{j=1}^{d} \sum_{k=1}^{d} \sum_{l=1}^{d} \frac{\partial^3 f(a_1,\ldots,a_d)}{\partial x_j \partial x_k \partial x_l}(x_j-a_j)(x_k-a_k)(x_l-a_l) + \cdots$$

For example, for a function $f(x, y)$ that depends on two variables, x and y, the Taylor series to second order about the point (a, b) is:

$$f(a,b)+(x-a)f_x(a,b)+(y-b)f_y(a,b)+\frac{1}{2!}\left((x-a)^2 f_{xx}(a,b)+2(x-a)(y-b)f_{xy}(a,b)+(y-b)^2 f_{yy}(a,b)\right)$$

where the subscripts denote the respective partial derivatives.

A second-order Taylor series expansion of a scalar-valued function of more than one variable can be written compactly as,

$$T(\mathbf{x}) = f(\mathbf{a}) + (\mathbf{x}-\mathbf{a})^{\mathsf{T}} Df(\mathbf{a}) + \frac{1}{2!}(\mathbf{x}-\mathbf{a})^{\mathsf{T}}\left\{D^2 f(\mathbf{a})\right\}(\mathbf{x}-\mathbf{a}) + \cdots,$$

where $Df(\mathbf{a})$ is the gradient of f evaluated at $\mathbf{x} = \mathbf{a}$ and $D^2 f(\mathbf{a})$ is the Hessian matrix. Applying the multi-index notation the Taylor series for several variables becomes,

$$T(\mathbf{x}) = \sum_{|\alpha| \geq 0} \frac{(\mathbf{x}-\mathbf{a})^{\alpha}}{\alpha!}\left(\partial^{\alpha} f\right)(\mathbf{a}),$$

which is to be understood as a still more abbreviated multi-index version of the first equation of this paragraph, with a full analogy to the single variable case.

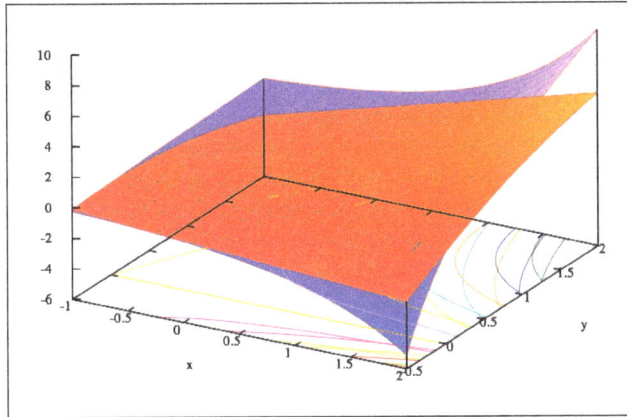

Second-order Taylor series approximation (in orange) of a function
$f(x,y) = e^x \ln(1+y)$ around the origin.

In order to compute a second-order Taylor series expansion around point $(a, b) = (0, 0)$ of the function,

$$f(x, y) = e^x \ln(1+y),$$

one first computes all the necessary partial derivatives:

$$f_x = e^x \ln(1+y)$$

$$f_y = \frac{e^x}{1+y}$$

$$f_{xx} = e^x \ln(1+y)$$

$$f_{yy} = -\frac{e^x}{(1+y)^2}$$

$$f_{xy} = f_{yx} = \frac{e^x}{1+y}.$$

Evaluating these derivatives at the origin gives the Taylor coefficients:

$$f_x(0,0) = 0$$
$$f_y(0,0) = 1$$
$$f_{xx}(0,0) = 0$$
$$f_{yy}(0,0) = -1$$
$$f_{xy}(0,0) = f_{yx}(0,0) = 1.$$

Substituting these values in to the general formula:

$$T(x,y) = f(a,b) + (x-a)f_x(a,b) + (y-b)f_y(a,b) + \frac{1}{2!}\left((x-a)^2 f_{xx}(a,b) + 2(x-a)(y-b)f_{xy}(a,b) + (y-b)^2 f_{yy}(a,b)\right) + \cdots$$

produces,

$$T(x,y) = 0 + 0(x-0) + 1(y-0) + \frac{1}{2}\left(0(x-0)^2 + 2(x-0)(y-0) + (-1)(y-0)^2\right) + \cdots$$

$$= y + xy - \frac{y^2}{2} + \cdots$$

Since $\log(1+y)$ is analytic in $|y| < 1$, we have:

$$e^x \log(1+y) = y + xy - \frac{y^2}{2} + \cdots, \qquad |y| < 1.$$

INTEGRAL CALCULUS

Integral calculus is the branch of mathematics in which the notion of an integral, its properties and methods of calculation are studied. Integral calculus is intimately related to differential calculus, and together with it constitutes the foundation of mathematical analysis. The origin of integral calculus goes back to the early period of development of mathematics and it is related to the method of exhaustion developed by the mathematicians of Ancient Greece. This method arose in the solution of problems on calculating areas of plane figures and surfaces, volumes of solid bodies, and in the solution of certain problems in statistics and hydrodynamics. It is based on the approximation of the objects under consideration by stepped figures or bodies, composed of simplest planar figures or special bodies (rectangles, parallelopipeds, cylinders, etc.). In this sense, the method of exhaustion can be regarded as an early method of integration. The greatest development of the method of exhaustion in the early period was obtained in the works of Eudoxus (4th century B.C.) and especially Archimedes (3rd century B.C.). Its subsequent application and perfection is associated with the names of several scholars of the 15th–17th centuries.

The fundamental concepts and theory of integral and differential calculus, primarily the relationship between differentiation and integration, as well as their application to the solution of applied problems, were developed in the works of P. de Fermat, I. Newton and G. Leibniz at the end of the 17th century. Their investigations were the beginning of an intensive development of mathematical analysis. The works of L. Euler, Jacob and Johann Bernoulli and J.L. Lagrange played an essential role in its creation in the 18th century. In the 19th century, in connection with the appearance of the notion of a limit, integral calculus achieved a logically complete form (in the works of A.L. Cauchy, B. Riemann and others). The development of the theory and methods of integral calculus took place at the end of 19th century and in the 20th century simultaneously with research into measure theory, which plays an essential role in integral calculus.

By means of integral calculus it became possible to solve by a unified method many theoretical and applied problems, both new ones which earlier had not been amenable to solution, and old ones that had previously required special artificial techniques. The basic notions of integral calculus are two closely related notions of the integral, namely the indefinite and the definite integral.

The indefinite integral of a given real-valued function on an interval on the real axis is defined as the collection of all its primitives on that interval, that is, functions whose derivatives are the given function. The indefinite integral of a function f is denoted by $\int f(x)\, dx$. If F is some primitive of f, then any other primitive of it has the form $F + C$, where C is an arbitrary constant; one therefore writes,

$$\int f(x)\, dx = F(x) + C \cdot$$

The operation of finding an indefinite integral is called integration. Integration is the operation inverse to that of differentiation:

$$\int F'(x)\, dx = F(x) + C, \qquad d\int f(x)\, dx = f(x)\, dx.$$

The operation of integration is linear: If on some interval the indefinite integrals,

$$\int f_1(x)dx \quad \text{and} \quad \int f_2(x)\, dx$$

exist, then for any real numbers λ_1 and λ_2, the following integral exists on this interval:

$$\int [\lambda_1 \, f_1(x) + \lambda_2 \, f_2(x)]\, dx$$

and equals,

$$\lambda_1 \int f_1(x)\, dx \ + \ \lambda_2 \int f_2(x)\, dx.$$

For indefinite integrals, the formula of integration by parts holds: If two functions u and v are differentiable on some interval and if the integral $\int v\, du$ exists, then so does the integral $\int u\, dv$, and the following formula holds:

$$\int u\, dv = uv - \int v\, du$$

The formula for change of variables holds: If for two functions f and ϕ defined on certain intervals, the composite function $f \circ \phi$ makes sense and the function ϕ is differentiable, then the integral,

$$\int f[\phi(t)] \, \phi'(t) \, dt$$

exists and equals:

$$\int f(x) \, dx.$$

A function that is continuous on some bounded interval has a primitive on it and hence an indefinite integral exists for it. The problem of actually finding the indefinite integral of a specified function is complicated by the fact that the indefinite integral of an elementary function is not an elementary function, in general. Many classes of functions are known for which it proves possible to express their indefinite integrals in terms of elementary functions. The simplest examples of these are integrals that are obtained from a table of derivatives of the basic elementary functions:

$$\int x^\alpha dx = \frac{x^{\alpha+1}}{\alpha+1} + C, \alpha \neq -1;$$

$$\int \frac{dx}{x} = \ln|x| + C;$$

$$\int a^x dx = \frac{a^x}{\ln \alpha} + C, \ \alpha > 0, \ \alpha \neq 1; \ \text{in particular,} \ \int e^x dx = e^x + C;$$

$$\int \sin x \, dx = -\cos x + C;$$

$$\int \cos x \, dx = \sin x + C;$$

$$\int \frac{dx}{\cos^2 x} = \tan x + C;$$

$$\int \frac{dx}{\sin^2 x} = -\cotan x + C;$$

$$\int \sinh x dx = \cosh x + C;$$

$$\int \cosh x \, dx = \sinh x + C;$$

$$\int \frac{dx}{\cosh^2 x} = \tanh x + C;$$

$$\int \frac{dx}{\sinh^2 x} = \cotanh x + C;$$

$$\int \frac{dx}{x^2 + \alpha^2} = \frac{1}{\alpha} \arctan \frac{x}{\alpha} + C = -\frac{1}{\alpha} \arccotan \frac{x}{\alpha} + C';$$

$$\int \frac{dx}{x^2 - \alpha^2} = \frac{1}{2\alpha} \ln \left| \frac{x-\alpha}{x+\alpha} \right| + C;$$

$$\int \frac{dx}{\sqrt{\alpha^2 - x^2}} = \arcsin \frac{x}{\alpha} + C = -\arccos \frac{x}{\alpha} + C', |x| < |\alpha|;$$

$$\int \frac{dx}{\sqrt{x^2 \pm \alpha^2}} = \ln |x + \sqrt{x^2 \pm \alpha^2}| + C \text{ (when } x^2 - \alpha^2 \text{ is under the square root, it is assumed that } |x| > |\alpha|).$$

If the denominator of the integrand vanishes at some point, then these formulas are valid only for those intervals inside which the denominator does not vanish.

The indefinite integral of a rational function over any interval on which the denominator does not vanish is a composition of rational functions, arctangents and natural logarithms. Finding the algebraic part of the indefinite integral of a rational function can be achieved by the Ostrogradski method. Integrals of the following types can be reduced by means of substitution and integration by parts to integration of rational functions:

$$\int R \left[x, \left(\frac{\alpha x + b}{cx + b} \right)^{r_1}, ..., \left(\frac{\alpha x + b}{cx + b} \right)^{r_m} \right] dx,$$

where $r_1, ..., r_m$ are rational numbers; integrals of the form,

$$\int R(x, \sqrt{ax^2 + bx + c}) dx$$

Certain cases of integrals of differential binomials; integrals of the form,

$$\int R(\sin x, \cos x) \, dx, \quad \int R(\sinh x, \cosh x) \, dx$$

(where $R(y_1, ..., y_n)$ are rational functions); the integrals,

$$\int e^{\alpha x} \cos \beta \, x dx, \quad \int e^{\alpha x} \sin \beta \, x dx,$$

$$\int x^n \cos \alpha \, x \, dx, \quad \int x^n \sin \alpha \, x \, dx,$$

$$\int x^n \arcsin x \, dx, \quad \int x^n \arccos x \, dx,$$

$$\int x^n \arctan x dx, \quad \int x^n \text{arccotan } x dx, \quad n = 0, 1, ...,$$

and many others. In contrast, for example, the integrals,

$$\int \frac{e^x}{x^n} \, dx, \quad \int \frac{\sin x}{x^n} \, dx, \quad \int \frac{\cos x}{x^n} \, dx, \quad n = 1, 2, ...,$$

cannot be expressed in terms of elementary functions.

The definite integral:

$$\int_{\alpha}^{b} f(x) \, dx$$

of a function f defined on an interval $[a, b]$ is the limit of integral sums of a specific type. If this limit exists, f is said to be Cauchy, Riemann, Lebesgue, etc. integrable.

The geometrical meaning of the integral is tied up with the notion of area: If the function $f \geq 0$ is continuous on the interval $[a, b]$, then the value of the integral,

$$\int\limits_{a}^{b} f(x)dx$$

is equal to the area of the curvilinear trapezium formed by the graph of the function, that is, the set whose boundary consists of the graph of f, the segment $[a, b]$ and the two segments on the lines $x = a$ and $x = b$ making the figure closed, which may degenerate to points.

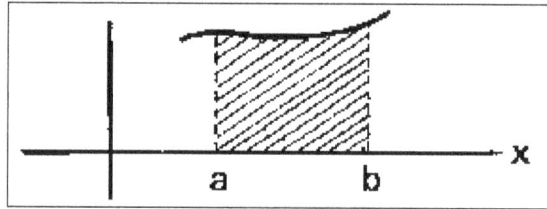

The calculation of many quantities encountered in practice reduces to the problem of calculating the limit of integral sums; in other words, finding a definite integral; for example, areas of figures and surfaces, volumes of bodies, work done by force, the coordinates of the centre of gravity, the values of the moments of inertia of various bodies, etc.

The definite integral is linear: If two functions f_1 and f_2 are integrable on an interval, then for any real numbers λ_1 and λ_2 the function:

$$\lambda_1 f_1 + \lambda_2 f_2$$

is also integrable on this interval and,

$$\int\limits_{a}^{b} \left[\lambda_1 f_1(x) + \lambda_2 f_2(x) \right] dx = \lambda_1 \int\limits_{a}^{b} f_1(x)dx + \lambda_2 \int\limits_{a}^{b} f_2(x)\, dx.$$

Integration of a function over an interval has the property of monotonicity: If the function f is integrable on the interval $[a, b]$ and if $[c, d] \subset [a, b]$, then f is integrable on $[c, d]$ as well. The integral is also additive with respect to the intervals over which the integration is carried out: If $a < c < b$ and the function f is integrable on the intervals $[a, c]$ and $[c, d]$, then it is integrable on $[a, b]$, and,

$$\int\limits_{a}^{b} f(x)\, dx = \int\limits_{a}^{c} f(x)\, dx + \int\limits_{c}^{b} f(x)\, dx.$$

If f and g are Riemann integrable, then their product is also Riemann integrable. If $f \geq g$ on $[a, b]$, then,

$$\int\limits_{a}^{b} f(x)\, dx \geq \int\limits_{a}^{b} g(x)\, dx.$$

If f is integrable on $[a, b]$, then the absolute value $|f|$ is also integrable on $[a, b]$ if $-\infty < a < b\infty$, and,

$$\left|\int_\alpha^b f(x)\,dx\right| \le \int_\alpha^b |f(x)|\,dx.$$

By definition one sets,

$$\int_\alpha^\alpha f(x)dx = 0 \text{ and } \int_b^\alpha f(x)dx = -\int_\alpha^b f(x)dx, \qquad a < b.$$

A mean-value theorem holds for integrals. For example, if f and g are Riemann integrable on an interval $[a, b]$, if $m \le f(x) \le M$, $x \in [a, b]$, and if g does not change sign on $[a, b]$, that is, it is either non-negative or non-positive throughout this interval, then there exists a number $m \le \mu \le M$ for which,

$$\int_\alpha^b f(x)g(x)\,dx = \mu \int_\alpha^b g(x)\,dx.$$

Under the additional hypothesis that f is continuous on $[a, b]$, there exists in (a, b) a point ξ for which,

$$\int_\alpha^b f(x)g(x)\,dx = f(\xi)\int_\alpha^b g(x)\,dx.$$

In particular, if $g(x) \equiv 1$, then,

$$\int_\alpha^b f(x)dx = f(\xi)(b - \alpha).$$

Integrals with a Variable Upper Limit

If a function f is Riemann integrable on an interval $[\alpha, b]$, then the function F defined by,

$$F(x) = \int_\alpha^x f(t)\,dt, \quad a \le x \le b,$$

is continuous on this interval. If, in addition, f is continuous at a point x_0, then F is differentiable at this point and $F'(x_0) = f(x_0)$. In other words, at the points of continuity of a function the following formula holds:

$$\frac{d}{dx}\int_\alpha^x f(t)\,dt = f(x).$$

Consequently, this formula holds for every Riemann-integrable function on an interval $[a, b]$, except perhaps at a set of points having Lebesgue measure zero, since if a function is Riemann integrable on some interval, then its set of points of discontinuity has measure zero. Thus, if the function f is continuous on $[a, b]$, then the function F defined by:

$$F(x) = \int_{\alpha}^{x} f(t)\, dt$$

is a primitive of f on this interval. This theorem shows that the operation of differentiation is inverse to that of taking the definite integral with a variable upper limit, and in this way a relationship is established between definite and indefinite integrals:

$$\int f(x)\, dx = \int_{\alpha}^{x} f(t)\, dt + C$$

The geometric meaning of this relationship is that the problem of finding the tangent to a curve and the calculation of the area of plane figures are inverse operations in the above sense.

The following Newton–Leibniz formula holds for any primitive F of an integrable function f on an interval $[a, b]$:

$$\int_{\alpha}^{b} f(x)\, dx = F(b) - F(a)$$

It shows that the definite integral of a continuous function over some interval is equal to the difference of the values at the end points of this interval of any primitive of it. This formula is sometimes taken as the definition of the definite integral. Then it is proved that the integral $\int_{\alpha}^{b} f(x)\, dx$ introduced in this way is equal to the limit of the corresponding integral sums.

For definite integrals, the formulas for change of variables and integration by parts hold. Suppose, for example, that the function f is continuous on the interval (a, b) and that ϕ is continuous together with its derivative ϕ' on the interval (α, β), where (α, β) is mapped by into (a, b) : $a < \phi(t) < b$ for $\alpha < t < \beta$, so that the composite $f \circ \phi$ is meaningful in (α, β). Then, for $\alpha_0, \beta_0 \in (\alpha, \beta)$, the following formulas for change of variables holds:

$$\int_{\phi(\alpha_0)}^{\phi(\beta_0)} f(x)\, dx = \int_{\alpha_0}^{\beta_0} f[\phi(t)]\, \phi'(t)\, dt.$$

The formula for integration by parts is:

$$\int_{a}^{b} u(x)\, dv(x) = u(x)v(x)\big|_{x=a}^{x=b} - \int_{a}^{b} v(x)\, du(x)$$

where the functions u and v have Riemann-integrable derivatives on $[a, b]$.

The Newton–Leibniz formula reduces the calculation of an indefinite integral to finding the values of its primitive. Since the problem of finding a primitive is intrinsically a difficult one, other methods of finding definite integrals are of great importance, among which one should mention the method of residues and the method of differentiation or integration with respect to the parameter of a parameter-dependent integral. Numerical methods for the approximate computation of integrals have also been developed.

Generalizing the notion of an integral to the case of unbounded functions and to the case of an unbounded interval leads to the notion of the improper integral, which is defined by yet one more limit transition.

The notions of the indefinite and the definite integral carry over to complex-valued functions. The representation of any holomorphic function of a complex variable in the form of a Cauchy integral over a contour played an important role in the development of the theory of analytic functions.

The generalization of the notion of the definite integral of a function of a single variable to the case of a function of several variables leads to the notion of a multiple integral.

For unbounded sets and unbounded functions of several variables, one is led to the notion of the improper integral, as in the one-dimensional case.

The extension of the practical applications of integral calculus necessitated the introduction of the notions of the curvilinear integral, i.e. the integral along a curve, the surface integral, i.e. the integral over a surface, and more generally, the integral over a manifold, which are reducible in some sense to a definite integral (the curvilinear integral reduces to an integral over an interval, the surface integral to an integral over a (plane) region, the integral over an n-dimensional manifold to an integral over an n-dimensional region). Integrals over manifolds, in particular curvilinear and surface integrals, play an important role in the integral calculus of functions of several variables; by this means a relationship is established between integration over a region and integration over its boundary or, in the general case, over a manifold and its boundary. This relationship is established by the Stokes formula.

Multiple, curvilinear and surface integrals find direct application in mathematical physics, particularly in field theory. Multiple integrals and concepts related to them are widely used in the solution of specific applied problems. The theory of cubature formulas has been developed for the numerical calculation of multiple integrals.

The theory and methods of integral calculus of real- or complex-valued functions of a finite number of real or complex variables carry over to more general objects. For example, the theory of integration of functions whose values lie in a normed linear space, functions defined on topological groups, generalized functions, and functions of an infinite number of variables (integrals over trajectories). Finally, a new direction in integral calculus is related to the emergence and development of constructive mathematics.

Integral calculus is applied in many branches of mathematics (in the theory of differential and integral equations, in probability theory and mathematical statistics, in the theory of optimal processes, etc.), and in applications of it.

References

- Differential-calculus: encyclopediaofmath.org, Retrieved 08 January, 2019

- Integral-calculus: encyclopediaofmath.org, Retrieved 16 June, 2019

- Michael Comenetz (2002). Calculus: The Elements. World Scientific. p. 159. ISBN 978-981-02-4904-5

- Hazewinkel, Michiel, ed. (2001) [1994], "Taylor series", Encyclopedia of Mathematics, Springer Science+Business Media B.V. / Kluwer Academic Publishers, ISBN 978-1-55608-010-4

Vector Calculus

Vector calculus, also known as vector analysis, is a mathematical branch that focuses on the differentiation and integration of vector fields. The main concepts used in vector calculus are Strokes' theorem, divergence theorem, curl, etc. The chapter closely examines the concepts of vector calculus to provide an extensive understanding of the subject.

Vector calculus is concerned with the differentiation and integration of vector fields, primarily in 3-dimensional Euclidean space. Here, the concept of a vector constitutes the mathematical abstraction of quantities that are characterized not only by a numerical value but also by a direction (for example, force, acceleration, velocity).

Vector Fields

In vector calculus, a vector field is an assignment of a vector to each point in a subset of Euclidean space. A vector field in the plane, for instance, can be visualized as a collection of arrows with a given magnitude and direction each attached to a point in the plane. Vector fields are often used to model the speed and direction of a moving fluid throughout space, for example, or the strength and direction of some force, such as the magnetic or gravitational force, as it changes from point to point.

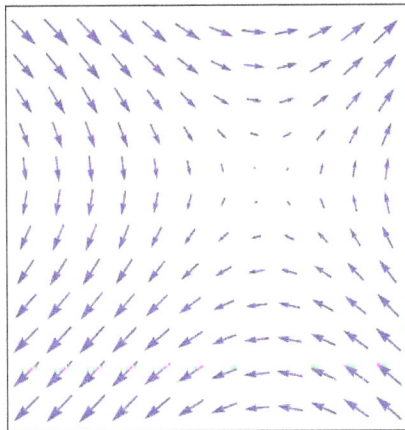

The elements of differential and integral calculus extend to vector fields in a natural way. When a vector field represents force, the line integral of a vector field represents the work done by a force moving along a path, and, under this interpretation, conservation of energy is exhibited as a special case of the fundamental theorem of calculus. Vector fields can be thought to represent the velocity of a moving flow in space, and this physical intuition leads to notions such as the divergence (the rate of change of volume of a flow) and curl (the rotation of a flow).

Gradient field: Vector fields can be constructed out of scalar fields using the gradient operator (denoted by the del: ∇). A vector field V defined on a set S is called a gradient field or a conservative field if there exists a real-valued function (a scalar field) f on S such that:

$$V = \nabla f = \left(\frac{\partial f}{\partial x_1}, \frac{\partial f}{\partial x_2}, \frac{\partial f}{\partial x_3}, \dots, \frac{\partial f}{\partial x_n} \right)$$

The associated flow is called the gradient flow.

Examples:

- A vector field for the movement of air on Earth will associate for every point on the surface of the Earth a vector with the wind speed and direction for that point.

- A gravitational field generated by any massive object is a vector field. For example, the gravitational field vectors for a spherically symmetric body would all point towards the sphere's center, with the magnitude of the vectors reducing as radial distance from the body increases.

- Magnetic field lines can be revealed using small iron filings.

- In the case of the velocity field of a moving fluid, a velocity vector is associated to each point in the fluid.

Conservative Vector Fields

A conservative vector field is a vector field which is the gradient of a function, known in this context as a scalar potential.

A conservative vector field is a vector field which is the gradient of a function, known in this context as a scalar potential. Conservative vector fields have the following property: The line integral from one point to another is independent of the choice of path connecting the two points; it is path-independent.

Conversely, path independence is equivalent to the vector field's being conservative. Conservative vector fields are also irrotational, meaning that (in three dimensions) they have vanishing curl. In fact, an irrotational vector field is necessarily conservative provided that a certain condition on the geometry of the domain holds: it must be simply connected.

A vector field v is said to be conservative if there exists a scalar field φ such that $v = \nabla \varphi$ Here denotes the gradient of φ. When the above equation holds, is called a scalar potential for v.

For any scalar field $\varphi : V \times \nabla \varphi = 0$. Therefore, the curl of a conservative vector field v is always 0. A vector field v whose curl is zero, is called irrotational.

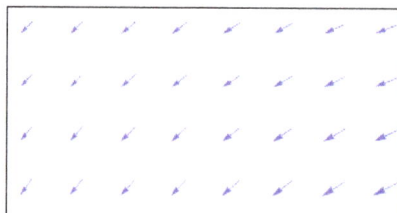

The above field $v(x, y, z) = (\dfrac{-y}{x^2 + y^2}, \dfrac{x}{x^2 + y^2}, 0)$ includes a vortex at its center, meaning it is non-irrotational; it is neither conservative, nor does it have path independence. However, any simply connected subset that excludes the vortex line $(0, 0, z)$ will have zero curl, $\nabla v = 0$. Such vortex-free regions are examples of irrotational vector fields.

Path Independence

A key property of a conservative vector field is that its integral along a path depends only on the endpoints of that path, not the particular route taken. Suppose that $S \subseteq \mathbb{R}^3$ is a region of three-dimensional space, and that P is a rectifiable path in S with start point A and end point B If $\mathbf{v} = \nabla \varphi$ is a conservative vector field, then the gradient theorem states that $\int_P \mathbf{v} \cdot d\mathbf{r} = \varphi(B) - \varphi(A)$ This holds as a consequence of the Chain Rule and the Fundamental Theorem of Calculus. An equivalent formulation of this is to say that $\oint \mathbf{v} \cdot d\mathbf{r} = 0$ for every closed loop in S.

DIFFERENTIATION OF VECTORS

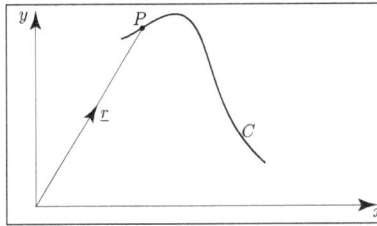

If r represents the position vector of an object which is moving along a curve C, then the position vector will be dependent upon the time, t. We write $\underline{r} = \underline{r}(t)$ to show the dependence upon time. Suppose that the object is at the point P, with position vector $\underline{r}$ at time t and at the point Q, with position vector $\underline{r}(t + \delta t)$, at the later time $t + \delta t$, as shown in figure.

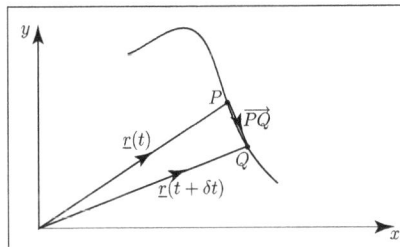

Then $\overrightarrow{PQ}$ represents the displacement vector of the object during the interval of time δt. The length of the displacement vector represents the distance travelled, and its direction gives the direction of motion. The average velocity during the time from t to $t + \delta t$ is defined as the displacement vector divided by the time interval δt, that is,

$$\text{average velocity} = \frac{\overrightarrow{PQ}}{\delta t} = \frac{\underline{r}(t + \delta t) - \underline{r}(t)}{\delta t}$$

If we now take the limit as the interval of time δt tends to zero then the expression on the right hand side is the derivative of $\underline{r}$ with respect to t. Not surprisingly we refer to this derivative as the instantaneous velocity, $\underline{v}$. By its very construction we see that the velocity vector is always tangential to the curve as the object moves along it. We have:

$$\underline{v} = \lim_{\delta t \to 0} \frac{\underline{r}(t+\delta t)-\underline{r}(t)}{\delta t} = \frac{d\underline{r}}{dt}$$

Now, since the x and y coordinates of the object depend upon time, we can write the position vector $\underline{r}$ in Cartesian coordinates as:

$$\underline{r}(t) = x(t)\underline{i} + y(t)\underline{j}$$

Therefore,

$$\underline{r}(t+\delta t) = x(t+\delta t)\underline{i} + y(t+\delta t)\underline{j}$$

so that,

$$\underline{v}(t) = \lim_{\delta t \to 0} \frac{x(t+\delta t)\underline{i} + y(t+\delta t)\underline{j} - x(t)\underline{i} - y(t)\underline{j}}{\delta t}$$

$$= \lim_{\delta t \to 0} \left\{ \frac{x(t+\delta t)-x(t)}{\delta t}\underline{i} + \frac{y(t+\delta t)-y(t)\underline{j}}{\delta t} \right\}$$

$$= \frac{dx}{dt}\underline{i} + \frac{dy}{dt}\underline{j}$$

This is often abbreviated to $\underline{v} = \underline{\dot{r}} = \dot{x}\underline{i} + \dot{y}\underline{j}$, using notation for derivatives with respect to time. So we see that the velocity vector is the derivative of the position vector with respect to time.

GRADIENT OF A SCALAR FIELD

The gradient of a scalar field is a vector field and whose magnitude is the rate of change and which points in the direction of the greatest rate of increase of the scalar field. If the vector is resolved, its components represent the rate of change of the scalar field with respect to each directional component. Hence for a two-dimensional scalar field $\phi(x,y)$.

$$\text{grad } \phi(x,y) = \nabla\phi(x,y) = \left(\frac{\partial}{\partial x}, \frac{\partial}{\partial y}\right)\phi = \left(\frac{\partial\phi}{\partial x}, \frac{\partial\phi}{\partial y}\right)$$

And for a three-dimensional scalar field $\phi(x,y,z)$

$$\text{grad } \phi(x,y,z) = \nabla\phi(x,y,z) = \left(\frac{\partial}{\partial x}, \frac{\partial}{\partial y}, \frac{\partial}{\partial z}\right)\phi = \left(\frac{\partial\phi}{\partial x}, \frac{\partial\phi}{\partial y}, \frac{\partial\phi}{\partial z}\right)$$

The gradient of a scalar field is the derivative of f in each direction. Note that the gradient of a scalar field is a vector field. An alternative notation is to use the del or nabla operator, $\nabla f = \text{grad } f$.

For a three dimensional scalar, its gradient is given by:

$$\text{grad } (V) = \underline{a}_n \frac{dV}{dn} = \nabla V$$

Gradient is a vector that represents both the magnitude and the direction of the maximum space rate of increase of a scalar:

$$dV = (\nabla V) \cdot dl, \text{ where } dl = \underline{a}i \cdot dl$$

In Cartesian:

$$\nabla \equiv \underline{a}_x \frac{\partial}{\partial x} + \underline{a}_y \frac{\partial}{\partial y} + \underline{a}_z \frac{\partial}{\partial z}$$

In Cylindrical:

$$\nabla \equiv \underline{a}_r \frac{\partial}{\partial r} + \underline{a}_\phi \frac{\partial}{r \cdot \partial \phi} + \underline{a}_z \frac{\partial}{\partial z}$$

In Spherical:

$$\nabla \equiv \underline{a}_R \frac{\partial}{\partial R} + \underline{a}_\emptyset \frac{\partial}{R \cdot \sin \theta \cdot \partial \phi}$$

Properties of gradient:

- We can change the vector field into a scalar field only if the given vector is differential. The given vector must be differential to apply the gradient phenomenon.

- The gradient of any scalar field shows its rate and direction of change in space.

Example: For the scalar field $\phi(x, y) = 3x + 5y$, calculate gradient of Ø.

Solution: Given scalar field $\phi(x, y) = 3x + 5y$

$$\nabla \phi(x, y) = \left(\frac{\partial \phi}{\partial x}, \frac{\partial \phi}{\partial y} \right) = (3, 5)$$

Example: For the scalar field $\phi(x, y) = x^4 yz$, calculate gradient of Ø.

Solution: Given scalar field $\phi(x, y) = x^4 yz$

$$\nabla \phi(x, y, z) = \left(\frac{\partial \phi}{ax}, \frac{\partial \phi}{\partial y}, \frac{\partial \phi}{\partial z} \right)$$

$$= \left(4x^3 yz, x^4 z, x^4 y \right)$$

Example: For the scalar field $\phi(x, y) = x^2 \sin 5y$, calculate gradient of ϕ.

Solution: Given scalar field $\phi(x, y) = x^2 \sin 5y$

$$\nabla \phi(x, y) = \left(\frac{\partial \phi}{\partial x}, \frac{\partial \phi}{\partial y} \right)$$
$$= \left(2x \sin(5y), 5x^2 \cos(5y) \right)$$

DIVERGENCE

The divergence is an operator, which takes in the vector-valued function defining this vector field, and outputs a scalar-valued function measuring the change in density of the fluid at each point.

This is the formula for divergence:

$$\text{div } \vec{v} = \nabla \cdot \vec{v} = \frac{\partial v^1}{\partial x} + \frac{\partial v^2}{\partial y} + \cdots$$

Here , $v^1, v^2, \ldots$ are the component functions of $\vec{v}$.

Changing Density in Fluid Flow

Take a look at the following vector field:

$$\vec{v}(x, y) = \left[\frac{2x - y}{y^2} \right]$$

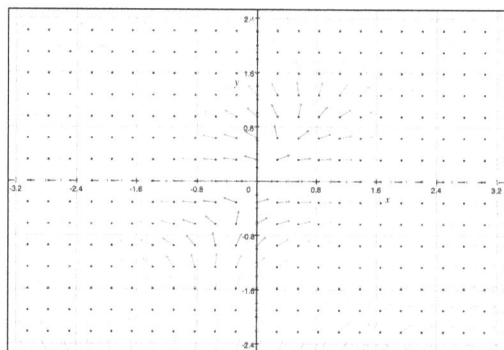

The inputs to $\vec{v}$ are points in two-dimensional space, (x,y), and the outputs are two-dimensional vectors, which in the vector field are attached to the corresponding point (x,y).

A nice way to think about vector fields is to imagine the fluid flow they could represent. Specifically, for each point (x,y) in two-dimensional space, imagine a particle sitting at (x,y) flowing in the direction of the vector attached to that point, $\vec{v}$(x,y). Moreover, suppose the speed of the particle's movement is determined by the length of that vector.

Notation and Formula for Divergence

The notation for divergence uses the same symbol "∇" which you may be familiar with from the gradient. As with the gradient, we think of this symbol loosely as representing a vector of partial derivative symbols.

$$\nabla = \begin{bmatrix} \dfrac{\partial}{\partial x} \\[2mm] \dfrac{\partial}{\partial y} \\[2mm] \vdots \end{bmatrix}$$

We write the divergence of a vector-valued function $\vec{v}(x, y, \dots)$ like this:

$$\nabla \cdot \vec{v} \leftarrow \text{Divergence of } \vec{v}$$

This is mildly nonsensical since ∇ isn't really a vector. Its entries are operators, not numbers. Nevertheless, using this dot product notation is super helpful for remembering how to compute divergence, just take a look:

$$\nabla \cdot \vec{v} = \begin{bmatrix} \dfrac{\partial}{\partial x} \\[2mm] \dfrac{\partial}{\partial y} \end{bmatrix} \cdot \begin{bmatrix} 2x - y \\ y^2 \end{bmatrix}$$

$$= \frac{\partial}{\partial x}(2x - y) + \frac{\partial}{\partial y}(y^2)$$

$$= 2 + 2y$$

More generally, the divergence can apply to vector-fields of any dimension. This means $\vec{v}$ can have any number of input variables, as long as its output has the same dimensions. Otherwise, it couldn't represent a vector field. If we write $\vec{v}$ component-wise like this:

$$\vec{v}(x_1, \dots, x_n) = \begin{bmatrix} v_1(x_1, \dots, x_n) \\ \vdots \\ v_n(x_1, \dots, x_n) \end{bmatrix}$$

Then the divergence of $\vec{v}$ looks like this:

$$\nabla \cdot \vec{v} = \begin{bmatrix} \dfrac{\partial}{\partial x_1} \\[2mm] \vdots \\[2mm] \dfrac{\partial}{\partial x_n} \end{bmatrix} \cdot \begin{bmatrix} v_1 \\ \vdots \\ v_n \end{bmatrix} = \frac{\partial v_1}{\partial x_1} + \cdots + \frac{\partial v_n}{\partial x_n}$$

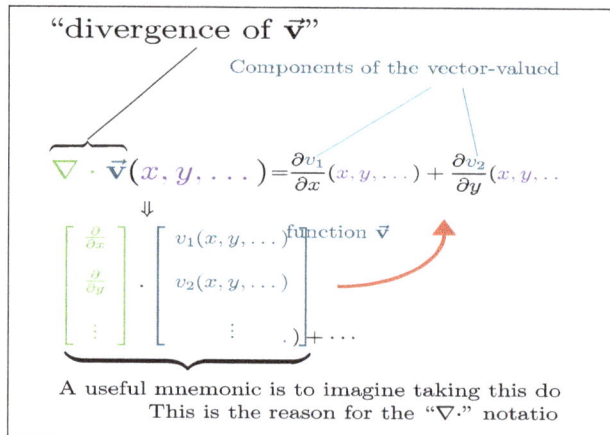

"divergence of $\vec{\mathbf{v}}$"

Components of the vector-valued

$$\nabla \cdot \vec{\mathbf{v}}(x, y, \dots) = \frac{\partial v_1}{\partial x}(x, y, \dots) + \frac{\partial v_2}{\partial y}(x, y, \dots)$$

$$\begin{bmatrix} \frac{\partial}{\partial x} \\ \frac{\partial}{\partial y} \\ \vdots \end{bmatrix} \cdot \begin{bmatrix} v_1(x, y, \dots) & \text{function } \vec{\mathbf{v}} \\ v_2(x, y, \dots) \\ \vdots \end{bmatrix} + \dots$$

A useful mnemonic is to imagine taking this do
This is the reason for the "$\nabla \cdot$" notatio

A useful mnemonic is to imagine taking this dot product.
This is the reason for the "∇" notation.

Interpretation of Divergence

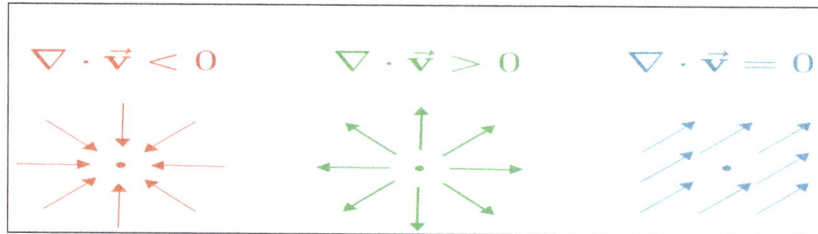

$\nabla \cdot \vec{\mathbf{v}} < 0$ $\nabla \cdot \vec{\mathbf{v}} > 0$ $\nabla \cdot \vec{\mathbf{v}} = 0$

Let's say you evaluate the divergence of a function $\vec{\mathbf{v}}$ at some point (x_0, y_0), and it comes out negative,

$$\nabla \cdot \vec{\mathbf{v}}(x_0, y_0) < 0$$

This means a fluid flowing along the vector field defined by $\vec{\mathbf{v}}$ would tend to become more dense at the point (x_0, y_0),

On the other hand, if the divergence at a point (x_0, y_0), is positive,

$$\nabla \cdot \vec{\mathbf{v}}(x_0, y_0) > 0$$

the fluid flowing along the vector field becomes less dense around (x_0, y_0). Finally, the concept of zero-divergence is very important in fluid dynamics and electrodynamics. It indicates that even though a fluid flows freely, its density stays constant. This is particularly handy when modeling incompressible fluids, such as water. In fact, the very idea that a fluid is incompressible can be tightly communicated with the following equation:

$$\nabla \cdot \vec{v} = 0$$

Such vector fields are called "divergence-free."

Sources and Sinks

Sometimes, for points with negative divergence, instead of thinking about a fluid getting more dense after a momentary fluid motion, some people imagine the fluid draining at that point while the fluid flows constantly.

As such, points of negative divergence are often called "sinks."

Likewise, instead of thinking of points with positive divergence as becoming less dense during a momentary motion, these points might be thought of as "sources" constantly generating more fluid particles.

CURL

The curl of a vector field, denoted curl (F) or $\nabla \times F$ (the notation used in this work), is defined as the vector field having magnitude equal to the maximum "circulation" at each point and to be oriented perpendicularly to this plane of circulation for each point. More precisely, the magnitude of $\nabla \times F$ is the limiting value of circulation per unit area. Written explicitly,

$$(\nabla \times F) \cdot \hat{n} \equiv \lim_{A \to 0} \frac{\oint_C F \cdot ds}{A},$$

where the right side is a line integral around an infinitesimal region of area A that is allowed to shrink to zero via a limiting process and $\hat{n}$ is the unit normal vector to this region. If $\nabla \times F = 0$ then the field is said to be an irrotational field. The symbol ∇ is variously known as "nabla" or "del".

The physical significance of the curl of a vector field is the amount of "rotation" or angular momentum of the contents of given region of space. It arises in fluid mechanics and elasticity theory. It is also fundamental in the theory of electromagnetism, where it arises in two of the four Maxwell equations,

$$\nabla \times E = -\frac{\partial B}{\partial t}$$

$$\nabla \times B = \mu_0 J + \varepsilon_0 \mu_0 \frac{\partial E}{\partial t},$$

where MKS units have been used here, E denotes the electric field, B is the magnetic field, μ_0 is a constant of proportionality known as the permeability of free space, J is the current density, and $\mathring{a}_0$ is another constant of proportionality known as the permittivity of free space. Together with the two other of the Maxwell equations, these formulas describe virtually all classical and relativistic properties of electromagnetism.

In Cartesian coordinates, the curl is defined by,

$$\nabla \times F = \left(\frac{\partial F_z}{\partial y} - \frac{\partial F_y}{\partial_z} \right) \hat{x} + \left(\frac{\partial F_x}{\partial z} \frac{\partial F_z}{\partial x} \right) \hat{y} + \left(\frac{\partial F_y}{\partial x} - \frac{\partial F_x}{\partial y} \right) \hat{z}.$$

This provides the motivation behind the adoption of the symbol $\nabla \times$ for the curl, since interpreting ∇ as the gradient operator $\nabla = (\partial / \partial x, \partial / \partial y, \partial / \partial z)$, the "cross product" of the gradient operator with   is given by,

$$\nabla \times \mathrm{F} = \begin{vmatrix} \hat{x} & \hat{y} & \hat{z} \\ \dfrac{\partial}{\partial x} & \dfrac{\partial}{\partial y} & \dfrac{\partial}{\partial z} \\ F_x & F_y & F_z \end{vmatrix},$$

which is precisely equation $\nabla \times \mathrm{F} = \left(\dfrac{\partial F_z}{\partial y} - \dfrac{\partial F_y}{\partial_z} \right) \hat{x} + \left(\dfrac{\partial F_x}{\partial z} \dfrac{\partial F_z}{\partial x} \right) \hat{y} + \left(\dfrac{\partial F_y}{\partial x} - \dfrac{\partial F_x}{\partial y} \right) \hat{z}$. A somewhat more elegant

formulation of the curl is given by the matrix operator equation:

$$\nabla \times \mathrm{F} = \begin{bmatrix} 0 & -\dfrac{\partial}{\partial z} & \dfrac{\partial}{\partial y} \\ \dfrac{\partial}{\partial z} & 0 & -\dfrac{\partial}{\partial x} \\ -\dfrac{\partial}{\partial y} & \dfrac{\partial}{\partial x} & 0 \end{bmatrix} \mathrm{F}$$

The curl can be similarly defined in arbitrary orthogonal curvilinear coordinates using,

$$\mathrm{F} \equiv F_1 \,\hat{\mathrm{u}}_1 + F_2 \,\hat{\mathrm{u}}_2 + F_3 \,\hat{\mathrm{u}}_3$$

and defining,

$$h_i \equiv \left| \frac{\partial \mathrm{r}}{\partial u_i} \right|,$$

as,

$$\nabla \times \mathrm{F} \equiv \frac{1}{h_1 \, h_2 \, h_3} \begin{vmatrix} h_1 \hat{u}_1 & h_2 \hat{u}_2 & h_3 \hat{u}_3 \\ \dfrac{\partial}{\partial u_1} & \dfrac{\partial}{\partial u_2} & \dfrac{\partial}{\partial u_3} \\ h_1 F_1 & h_2 F_2 & h_3 F_3 \end{vmatrix}$$

$$= \frac{1}{h_2 \, h_3} \left[\frac{\partial}{\partial u_2} (h_3 \, F_3) - \frac{\partial}{\partial u_3} (h_2 F_2) \right] \hat{u}_1 + \frac{1}{h_1 \, h_3} \left[$$

$$\frac{\partial}{\partial u_3} (h_1 \, F_1) - \frac{\partial}{\partial_{u1}} (h_3 \, F_3) \right] \hat{u}_2 + \frac{1}{h_1 \, h_2} \left[\frac{\partial}{\partial_{u1}} (h_2 \, F_2) - \frac{\partial}{\partial_{u2}} (h_1 \, F_1) \right] \hat{u}_3.$$

The curl can be generalized from a vector field to a tensor field as:

$$(\nabla \times A)^a = \varepsilon^{a\mu\nu} \, a_{\nu;\mu},$$

where ε_{ijk} is the permutation tensor and ";" denotes a covariant derivative.

VECTOR INTEGRATION

In ordinary calculus we compute integrals of real functions of a real variable; that is, we compute integrals of functions of the type:

$$y = f(x)$$

where x and y are real numbers.

In vector analysis we compute integrals of vector functions of a real variable; that is we compute integrals of functions of the type:

$$f(t) = f_1(t)i + f_2(t)j + f_3(t)k$$

or equivalently,

$$\overline{f(t)} = \begin{bmatrix} f_1(t) \\ f_2(t) \\ f_3(t) \end{bmatrix}$$

where $f_1(t)$, $f_2(t)$, and $f_3(t)$ are real functions of the real variable t.

Line Integrals

In mathematics, a line integral is an integral where the function to be integrated is evaluated along a curve. The terms path integral, curve integral, and curvilinear integral are also used; contour integral as well, although that is typically reserved for line integrals in the complex plane.

The function to be integrated may be a scalar field or a vector field. The value of the line integral is the sum of values of the field at all points on the curve, weighted by some scalar function on the curve (commonly arc length or, for a vector field, the scalar product of the vector field with a differential vector in the curve). This weighting distinguishes the line integral from simpler integrals defined on intervals. Many simple formulae in physics (for example, $W = F \cdot s$) have natural continuous analogs in terms of line integrals ($W = \int_C F \cdot ds$). The line integral finds the work done on an object moving through an electric or gravitational field, for example.

Vector Calculus

In qualitative terms, a line integral in vector calculus can be thought of as a measure of the total effect of a given tensor field along a given curve. For example, the line integral over a scalar field (rank 0 tensor) can be interpreted as the area under the field carved out by a particular curve. This can be visualized as the surface created by $z = f(x,y)$ and a curve C in the x-y plane. The line integral of f would be the area of the "curtain" created when the points of the surface that are directly over C are carved out.

Line Integral of a Scalar Field

For some scalar field $f : \mathbb{U} \subseteq \mathbb{R}^n \to \mathbb{R}$, the line integral along a piecewise smooth curve $\mathcal{C} \subset \mathbb{U}$ is defined as:

$$\int_{\mathcal{C}} f(\mathbf{r}) ds = \int_a^b f\big(\mathbf{r}(t)\big) | \mathbf{r}'(t) | dt.$$

where $\mathbf{r} : [a,b] \to \mathcal{C}$ is an arbitrary bijective parametrization of the curve $\mathcal{C}$ such that $\mathbf{r}(a)$ and $\mathbf{r}(b)$ give the endpoints of $\mathcal{C}$ and $a < b$. Here, and in the rest of the part, the absolute value bars denote the standard (euclidean) norm of a vector.

The function f is called the integrand, the curve $\mathcal{C}$ is the domain of integration, and the symbol ds may be intuitively interpreted as an elementary arc length. Line integrals of scalar fields over a curve $\mathcal{C}$ do not depend on the chosen parametrization $\mathbf{r}$ of $\mathcal{C}$.

Geometrically, when the scalar field f is defined over a plane $(n = 2)$, its graph is a surface $z = f(x,y)$ in space, and the line integral gives the (signed) cross-sectional area bounded by the curve $\mathcal{C}$ and the graph of f.

Derivation

For a line integral over a scalar field, the integral can be constructed from a Riemann sum using the above definitions of f, C and a parametrization $\mathbf{r}$ of C. This can be done by partitioning the interval $[a,b]$ into n sub-intervals $[t_{i-1}, t_i]$ of length $\Delta t = (b - a)/n$, then $\mathbf{r}(t_i)$ denotes some point, call it a sample point, on the curve C. We can use the set of sample points $\{\mathbf{r}(t_i) : 1 \leq i \leq n\}$ to approximate the curve C by a polygonal path by introducing a straight line piece between each of the sample points $\mathbf{r}(t_{i-1})$ and $\mathbf{r}(t_i)$. We then label the distance between each of the sample points on the curve as Δs_i. The product of $f(\mathbf{r}(t_i))$ and Δs_i can be associated with the signed area of a rectangle with a height and width of $f(\mathbf{r}(t_i))$ and Δs_i respectively. Taking the limit of the sum of the terms as the length of the partitions approaches zero gives us:

$$I = \lim_{\Delta s_i \to 0} \sum_{i=1}^{n} f(\mathbf{r}(t_i)) \Delta s_i.$$

We note that, by the mean value theorem, the distance between subsequent points on the curve, is

$$\Delta s_i = | \mathbf{r}(t_i + \Delta t) - \mathbf{r}(t_i) | \approx | \mathbf{r}'(t_i) | \Delta t.$$

Substituting this in the above Riemann sum yields,

$$I = \lim_{\Delta t \to 0} \sum_{i=1}^{n} f(\mathbf{r}(t_i)) | \mathbf{r}'(t_i) | \Delta t$$

which is the Riemann sum for the integral,

$$I = \int_a^b f(\mathbf{r}(t)) | \mathbf{r}'(t) | dt.$$

Line Integral of a Vector Field

For a vector field $F : U \subseteq \mathbb{R}^n \to \mathbb{R}^n$, the line integral along a piecewise smooth curve $C \subset U$, in the direction of $\mathbf{r}$, is defined as,

$$\int_C \mathbf{F}(\mathbf{r}) \cdot d\mathbf{r} = \int_a^b \mathbf{F}(\mathbf{r}(t)) \cdot \mathbf{r}'(t) dt.$$

where $\cdot$ is the dot product and r: $[a, b] \to C$ is a bijective parametrization of the curve C such that r(a) and r(b) give the endpoints of C.

A line integral of a scalar field is thus a line integral of a vector field where the vectors are always tangential to the line.

Line integrals of vector fields are independent of the parametrization $\mathbf{r}$ in absolute value, but they do depend on its orientation. Specifically, a reversal in the orientation of the parametrization changes the sign of the line integral.

From the viewpoint of differential geometry, the line integral of a vector field along a curve is the integral of the corresponding 1-form under the musical isomorphism (which takes the vector field to the corresponding covector field) over the curve considered as an immersed 1-manifold.

Derivation

The line integral of a vector field can be derived in a manner very similar to the case of a scalar field, but this time with the inclusion of a dot product. Again using the above definitions of F, C and its parametrization r(t), we construct the integral from a Riemann sum. We partition the interval $[a,b]$ (which is the range of the values of the parameter t) into n intervals of length $\Delta t = (b - a)/n$. Letting t_i be the ith point on $[a,b]$, then r(t_i) gives us the position of the ith point on the curve. However, instead of calculating up the distances between subsequent points, we need to calculate their displacement vectors, $\Delta \mathbf{r}_i$. As before, evaluating F at all the points on the curve and taking the dot product with each displacement vector gives us the infinitesimal contribution of each partition of F on C. Letting the size of the partitions go to zero gives us a sum,

$$I = \lim_{\Delta t \to 0} \sum_{i=1}^{n} \mathbf{F}(\mathbf{r}(t_i)) \cdot \Delta \mathbf{r}_i$$

By the mean value theorem, we see that the displacement vector between adjacent points on the curve is,

$$\Delta \mathbf{r}_i = \mathbf{r}(t_i + \Delta t) - \mathbf{r}(t_i) \approx \mathbf{r}'(t_i) \Delta t$$

Substituting this in the above Riemann sum yields,

$$I = \lim_{\Delta t \to 0} \sum_{i=1}^{n} \mathbf{F}(\mathbf{r}(t_i)) \cdot \mathbf{r}'(t_i) \Delta t$$

which is the Riemann sum for the integral defined above.

Path Independence

If a vector field F is the gradient of a scalar field G (i.e. if F is conservative), that is,

$$\nabla G = \mathbf{F},$$

then the derivative of the composition of G and $\mathbf{r}(t)$ is,

$$\frac{dG(\mathbf{r}(t))}{dt} = \nabla G(\mathbf{r}(t)) \cdot \mathbf{r}'(t) = \mathbf{F}(\mathbf{r}(t)) \cdot \mathbf{r}'(t)$$

which happens to be the integrand for the line integral of F on r(t). It follows, given a path C, that

$$\int_C \mathbf{F}(\mathbf{r}) \cdot d\mathbf{r} = \int_a^b \mathbf{F}(\mathbf{r}(t)) \cdot \mathbf{r}'(t) dt = \int_a^b \frac{dG(\mathbf{r}(t))}{dt} dt = G(\mathbf{r}(b)) - G(\mathbf{r}(a)).$$

In other words, the integral of **F** over C depends solely on the values of G at the points $\mathbf{r}(b)$ and $\mathbf{r}(a)$ and is thus independent of the path between them. For this reason, a line integral of a conservative vector field is called *path independent*.

Applications

The line integral has many uses in physics. For example, the work done on a particle traveling on a curve C inside a force field represented as a vector field **F** is the line integral of **F** on C.

Flow across a Curve

For a vector field $F : U \subseteq R^2 \rightarrow R^2$, such as $F(x, y) = (P(x, y), Q(x, y))$ the line integral across a piecewise smooth curve $C \subset U$, is defined as

$$\int_C \mathbf{F}(\mathbf{r}) \cdot d\mathbf{r}' = \int_a^b -Q(x, y) dx + P(x, y) dy = \int_a^b (P(\mathbf{r}(t)), Q(\mathbf{r}(t))) \cdot (r_2'(t), -r_1'(t)) dt.$$

where $\cdot$ is the dot product and $r : [a, b] \rightarrow C$, $r(t) = (r_1(t), r_2(t))$ is a bijective parametrization of the curve C such that r(a) and r(b) give the endpoints of C.

Complex Line Integral

In complex analysis, the line integral is defined in terms of multiplication and addition of complex numbers. Suppose U is an open subset of the complex plane C, $f : U \rightarrow C$ is a function, and $L \subset U$ is a curve of finite length, parametrized by $\gamma : [a, b] \rightarrow L$, where $\gamma(t) = x(t) + iy(t)$. The line integral:

$$\int_L f(z) dz$$

may be defined by subdividing the interval $[a, b]$ into $a = t_0 < t_1 < ... < t_n = b$ and considering the expression:

$$\sum_{k=1}^n f(\gamma(t_k))[\gamma(t_k) - \gamma(t_{k-1})] = \sum_{k=1}^n f(\gamma_k) \Delta \gamma_k.$$

The integral is then the limit of this Riemann sum as the lengths of the subdivision intervals approach zero.

If the parametrization γ is continuously differentiable, the line integral can be evaluated as an integral of a function of a real variable:

$$\int_L f(z)dz = \int_a^b f(\gamma(t))\gamma'(t)dt.$$

When L is a closed curve, that is, its initial and final points coincide, the notation,

$$\oint_L f(z)dz$$

is often used for the line integral of f along L. A closed curve line integral is sometimes referred to as a cyclic integral in engineering applications.

The line integral with respect to the conjugate complex differential $\overline{dz}$ is defined to be,

$$\int_L f\,\overline{dz} := \overline{\int_L \overline{f}\,dz} = \int_a^b f(\gamma(t))\overline{\gamma'(t)}dt.$$

The line integrals of complex functions can be evaluated using a number of techniques: the integral may be split into real and imaginary parts reducing the problem to that of evaluating two real-valued line integrals, the Cauchy integral formula may be used in other rcumstances. If the line integral is a closed curve in a region where the function is analytic and containing no singularities, then the value of the integral is simply zero; this is a consequence of the Cauchy integral theorem. The residue theorem allows contour integrals to be used in the complex plane to find integrals of real-valued functions of a real variable.

Consider the function $f(z)=1/z$, and let the contour L be the unit circle about 0, parametrized by $z(t)=e^{it}$ with t in $[0, 2\pi]$ (which generates the circle counterclockwise). Substituting, we find,

$$\oint_L f(z)dz = \int_0^{2\pi} \frac{1}{e^{it}}ie^{it}\,dt = i\int_0^{2\pi} e^{-it}e^{it}\,dt = i\int_0^{2\pi} dt$$
$$= i(2\pi - 0) = 2\pi i.$$

Here we have used the fact that any complex number z can be written as re^{it} where r is the modulus of z. On the unit circle this is fixed to 1, so the only variable left is the angle, which is denoted by t. This answer can be also verified by Cauchy's integral formula.

Relation between the line integral of a vector field and the complex line integral:

Viewing complex numbers as 2-dimensional vectors, the line integral of a 2-dimensional vector field corresponds to the real part of the line integral of the conjugate of the corresponding complex function of a complex variable. More specifically, if $\mathbf{r}(t) = (x(t), y(t))$ is a parameterization of L and $f(z) = u(z) + iv(z)$, then:

$$\int_L \overline{f(z)}dz = \int_L \overline{f}\,dx + i\int_L \overline{f}\,dy = \int_L (u,v)\cdot d\mathbf{r} + i\int_L (-v,u)\cdot d\mathbf{r},$$

provided that both integrals on the right hand side exist, and that the parametrization γ of L has the same orientation as $\mathbf{r}$ (just expand the Riemann sum for the lefthand integral and take the limit).

By Green's theorem, the area of a region enclosed by a smooth, closed, positively oriented curve L is given by the integral,

$$\frac{1}{2i}\int_{L}\bar{z}\,dz$$

This fact is used, for example, in the proof of the area theorem.

Due to the Cauchy–Riemann equations the curl of the vector field corresponding to the conjugate of a holomorphic function is zero. This relates through Stokes' theorem both types of line integral being zero.

Geometric Line Integral

A curve γ embedded in some vectorspace V is a pure geometric object, hence it does not need a specific coordinate representation to exist. Let F be a one form (for example force) or a linear function on V that takes vectors v from V and maps them into the reals. Then one can integrate F along a one dimensional object like that curve γ,

$$\int_{\gamma}F$$

Intuitively one feeds this linear function with all the infinitesimal tangent vectors that are attached at each point of γ. In coordinates x^i the one form has the representation (Note that one sums over repeated indices),

$$\int_{\gamma}F = \int_{\gamma}f_i(x)dx^i$$

If one parametrizes the curve by some parameter $t \in [0,1]$, by pullback one arrives at the well known line integral form,

$$\int_0^1 f_i(x(t))\frac{\partial x^i}{\partial t}\,dt$$

Note that we made no use of any scalar product hence it is possible to define line integrals without the use of a metric. Of course with a scalar product in hand, the metric induces a map that identifies vectors with 1 forms and one would arrive at the usual definition of line integrals from vector calculus.

$$\int_0^1 \langle \mathbf{F}, \frac{\partial \mathbf{x}}{\partial t}\rangle dt$$

where $\mathbf{F}$ is the vector such that.

Surface Integrals

In mathematics, a surface integral is a generalization of multiple integrals to integration over surfaces. It can be thought of as the double integral analogue of the line integral. Given a surface, one may integrate over its scalar fields (that is, functions which return scalars as values), and vector fields (that is, functions which return vectors as values).

Surface integrals have applications in physics, particularly with the theories of classical electromagnetism.

Surface Integrals of Scalar Fields

To find an explicit formula for the surface integral, we need to parameterize the surface of interest, S, by considering a system of curvilinear coordinates on S, like the latitude and longitude on a sphere. Let such a parameterization be $\mathbf{x}(s, t)$, where (s, t) varies in some region T in the plane. Then, the surface integral is given by,

$$\iint_S f\, dS = \iint_T f(\mathbf{x}(s,t)) \left\| \frac{\partial \mathbf{x}}{\partial s} \times \frac{\partial \mathbf{x}}{\partial t} \right\| ds\, dt$$

where the expression between bars on the right-hand side is the magnitude of the cross product of the partial derivatives of $\mathbf{x}(s, t)$, and is known as the surface element. The surface integral can also be expressed in the equivalent form,

$$\iint_S f\, d\Sigma = \iint_T f(\mathbf{x}(s,t)) \sqrt{g}\, ds\, dt$$

where g is the determinant of the first fundamental form of the surface mapping x (s, t).

For example, if we want to find the surface area of the graph of some scalar function, say $z = f(x, y)$, we have

$$A = \iint_S d\Sigma = \iint_T \left\| \frac{\partial \mathbf{r}}{\partial x} \times \frac{\partial \mathbf{r}}{\partial y} \right\| dx\, dy$$

where $\mathbf{r} = (x, y, z) = (x, y, f(x, y))$. So that $\dfrac{\partial \mathbf{r}}{\partial x} = (1, 0, f_x(x, y))$, and $\dfrac{\partial \mathbf{r}}{\partial y} = (0, 1, f_y(x, y))$. So,

$$A = \iint_T \left\| \left(1, 0, \frac{\partial f}{\partial x}\right) \times \left(0, 1, \frac{\partial f}{\partial y}\right) \right\| dx\, dy$$

$$= \iint_T \left\| \left(-\frac{\partial f}{\partial x}, -\frac{\partial f}{\partial y}, 1\right) \right\| dx\, dy$$

$$= \iint_T \sqrt{\left(\frac{\partial f}{\partial x}\right)^2 + \left(\frac{\partial f}{\partial y}\right)^2 + 1}\, dx\, dy$$

which is the standard formula for the area of a surface described this way. One can recognize the vector in the second line above as the normal vector to the surface.

Note that because of the presence of the cross product, the above formulas only work for surfaces embedded in three-dimensional space.

This can be seen as integrating a Riemannian volume form on the parameterized surface, where the metric tensor is given by the first fundamental form of the surface.

Surface Integrals of Vector Fields

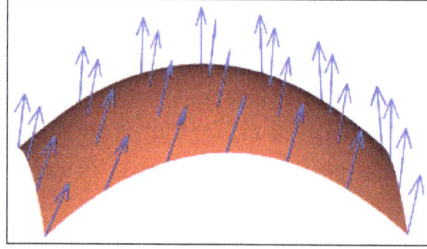

A vector field on a surface.

Consider a vector field v on S, that is, for each x in S, v(x) is a vector. The surface integral can be defined component-wise according to the definition of the surface integral of a scalar field; the result is a vector. This applies for example in the expression of the electric field at some fixed point due to an electrically charged surface, or the gravity at some fixed point due to a sheet of material.

Alternatively, if we integrate the normal component of the vector field, the result is a scalar. Imagine that we have a fluid flowing through S, such that v(x) determines the velocity of the fluid at x. The flux is defined as the quantity of fluid flowing through S per unit time.

This illustration implies that if the vector field is tangent to S at each point, then the flux is zero because the fluid just flows in parallel to S, and neither in nor out. This also implies that if v does not just flow along S, that is, if **v** has both a tangential and a normal component, then only the normal component contributes to the flux. Based on this reasoning, to find the flux, we need to take the dot product of **v** with the unit surface normal **n** to S at each point, which will give us a scalar field, and integrate the obtained field as above. We find the formula:

$$\iint_S \mathbf{v}\cdot d\Sigma = \iint_S (\mathbf{v}\cdot\mathbf{n})d\Sigma$$

$$= \iint_T \left(\mathbf{v}(\mathbf{x}(s,t))\cdot \frac{\left(\frac{\partial\mathbf{x}}{\partial s}\times\frac{\partial\mathbf{x}}{\partial t}\right)}{\left\|\left(\frac{\partial\mathbf{x}}{\partial s}\times\frac{\partial\mathbf{x}}{\partial t}\right)\right\|} \right) \left\|\left(\frac{\partial\mathbf{x}}{\partial s}\times\frac{\partial\mathbf{x}}{\partial t}\right)\right\| ds\,dt$$

$$= \iint_T \mathbf{v}(\mathbf{x}(s,t))\cdot\left(\frac{\partial\mathbf{x}}{\partial s}\times\frac{\partial\mathbf{x}}{\partial t}\right) ds\,dt.$$

The cross product on the right-hand side of this expression is a (not necessarily unital) surface normal determined by the parametrization.

This formula *defines* the integral on the left (note the dot and the vector notation for the surface element).

We may also interpret this as a special case of integrating 2-forms, where we identify the vector field with a 1-form, and then integrate its Hodge dual over the surface. This is equivalent to integrating $\langle \mathbf{v}, \mathbf{n} \rangle \, \mathrm{d}\Sigma$ over the immersed surface, where $\mathrm{d}\Sigma$ is the induced volume form on the surface, obtained by interior multiplication of the Riemannian metric of the ambient space with the outward normal of the surface.

Surface Integrals of Differential 2-forms

Let,

$$f = f_z \, \mathrm{d}x \wedge \mathrm{d}y + f_x \, \mathrm{d}y \wedge \mathrm{d}z + f_y \, \mathrm{d}z \wedge \mathrm{d}x$$

be a differential 2-form defined on the surface S, and let,

$$\mathbf{x}(s,t) = (x(s,t), y(s,t), z(s,t))$$

be an orientation preserving parametrization of S with (s,t) in D. Changing coordinates from (x,y) to (s,t), the differential forms transform as:

$$\mathrm{d}x = \frac{\partial x}{\partial s} \, \mathrm{d}s + \frac{\partial x}{\partial t} \, \mathrm{d}t$$

$$\mathrm{d}y = \frac{\partial y}{\partial s} \, \mathrm{d}s + \frac{\partial y}{\partial t} \, \mathrm{d}t$$

So $\mathrm{d}x \wedge \mathrm{d}y$ transforms to $\dfrac{\partial(x,y)}{\partial(s,t)} \, \mathrm{d}s \wedge \mathrm{d}t$, where $\dfrac{\partial(x,y)}{\partial(s,t)}$ denotes the determinant of the Jacobian of the transition function from (s,t) to (x,y). The transformation of the other forms are similar.

Then, the surface integral of f on S is given by:

$$\iint_D \left[f_z(\mathbf{x}(s,t)) \frac{\partial(x,y)}{\partial(s,t)} + f_x(\mathbf{x}(s,t)) \frac{\partial(y,z)}{\partial(s,t)} + f_y(\mathbf{x}(s,t)) \frac{\partial(z,x)}{\partial(s,t)} \right] \mathrm{d}s \, \mathrm{d}t$$

where,

$$\frac{\partial \mathbf{x}}{\partial s} \times \frac{\partial \mathbf{x}}{\partial t} = \left(\frac{\partial(y,z)}{\partial(s,t)}, \frac{\partial(z,x)}{\partial(s,t)}, \frac{\partial(x,y)}{\partial(s,t)} \right)$$

is the surface element normal to S.

Let us note that the surface integral of this 2-form is the same as the surface integral of the vector field which has as components f_x, f_y and f_z.

Volume Integrals

In mathematics—in particular, in multivariable calculus—a volume integral refers to an integral over a 3-dimensional domain, that is, it is a special case of multiple integrals. Volume integrals are especially important in physics for many applications, for example, to calculate flux densities.

In Coordinates

It can also mean a triple integral within a region $D \subset \mathbb{R}^3$ of a function $f(x, y, z)$, and is usually written as:

$$\iiint_D f(x, y, z)dx\, dy\, dz.$$

A volume integral in cylindrical coordinates is

$$\iiint_D f(\rho, \varphi, z)\rho d\rho\, d\varphi\, dz,$$

and a volume integral in spherical coordinates (using the ISO convention for angles with   as the azimuth and θ measured from the polar axis has the form

$$\iiint_D f(r, \theta, \varphi)r^2 \sin\theta\, dr\, d\theta\, d\varphi.$$

Example of Volume Integrals

Integrating the function $f(x, y, z) = 1$ over a unit cube yields the following result:

$$\int_0^1\int_0^1\int_0^1 1dx\,dy\,dz = \int_0^1\int_0^1 (1-0)dy\,dz = \int_0^1 (1-0)dz = 1-0 = 1$$

So the volume of the unit cube is 1 as expected. This is rather trivial however, and a volume integral is far more powerful. For instance if we have a scalar density function on the unit cube then the volume integral will give the total mass of the cube. For example for density function:

$$\begin{cases} f : \mathbb{R}^3 \to \mathbb{R} \\ (x, y, z) \mapsto x + y + z \end{cases}$$

the total mass of the cube is:

$$\int_0^1\int_0^1\int_0^1 (x+y+z)dx\,dy\,dz = \int_0^1\int_0^1 \left(\frac{1}{2}+y+z\right)dy\,dz = \int_0^1 (1+z)dz = \frac{3}{2}$$

STOKES' THEOREM

In vector calculus, and more generally differential geometry, Stokes' theorem (sometimes spelled Stokes's theorem, and also called the generalized Stokes theorem or the Stokes–Cartan theorem) is a statement about the integration of differential forms on manifolds, which both simplifies and generalizes several theorems from vector calculus. Stokes' theorem says that the integral of a differential form ω over the boundary of some orientable manifold Ω is equal to the integral of its exterior derivative $d\omega$ over the whole of Ω, i.e.,

$$\int_{\partial\Omega} \omega = \int_{\Omega} d\omega.$$

Stokes' theorem was formulated in its modern form by Élie Cartan in 1945, following earlier work on the generalization of the theorems of vector calculus by Vito Volterra, Édouard Goursat, and Henri Poincaré.

This modern form of Stokes' theorem is a vast generalization of a classical result that Lord Kelvin communicated to George Stokes in a letter dated July 2, 1850. Stokes set the theorem as a question on the 1854 Smith's Prize exam, which led to the result bearing his name. It was first published by Hermann Hankel in 1861. This classical Kelvin–Stokes theorem relates the surface integral of the curl of a vector field F over a surface in Euclidean three-space to the line integral of the vector field over its boundary: Let γ: $[a, b] \to$ R² be a piecewise smooth Jordan plane curve. The Jordan curve theorem implies that γ divides R² into two components, a compact one and another that is non-compact. Let D denote the compact part that is bounded by γ and suppose ψ: $D \to$ R³ is smooth, with $S := \psi(D)$. If Γ is the space curve defined by $\Gamma(t) = \psi(\gamma(t))$ and F is a smooth vector field on R³, then:

$$\oint_{\Gamma} \mathbf{F} \cdot d\Gamma = \iint_{S} \nabla \times \mathbf{F} \cdot d\mathbf{S}$$

This classical statement, along with the classical divergence theorem, the fundamental theorem of calculus, and Green's theorem are simply special cases of the general formulation stated above.

The fundamental theorem of calculus states that the integral of a function f over the interval $[a, b]$ can be calculated by finding an antiderivative F of f:

$$\int_{a}^{b} f(x)dx = F(b) - F(a).$$

Stokes' theorem is a vast generalization of this theorem in the following sense:

- By the choice of F, $\dfrac{dF}{dx} = f(x)$. In the parlance of differential forms, this is saying that $f(x)\,dx$ is the exterior derivative of the 0-form, i.e. function, F: in other words, that $dF = f\,dx$. The general Stokes theorem applies to higher differential forms ω instead of just 0-forms such as F.

- A closed interval $[a, b]$ is a simple example of a one-dimensional manifold with boundary. Its boundary is the set consisting of the two points a and b. Integrating f over the interval may be generalized to integrating forms on a higher-dimensional manifold. Two technical conditions are needed: the manifold has to be orientable, and the form has to be compactly supported in order to give a well-defined integral.

- The two points a and b form the boundary of the closed interval. More generally, Stokes' theorem applies to oriented manifolds M with boundary. The boundary ∂M of M is itself a manifold and inherits a natural orientation from that of M. For example, the natural orientation of the interval gives an orientation of the two boundary points. Intuitively, a inherits the opposite orientation as b, as they are at opposite ends of the interval. So, "integrating" F over two boundary points a, b is taking the difference $F(b) - F(a)$.

In even simpler terms, one can consider the points as boundaries of curves, that is as 0-dimensional boundaries of 1-dimensional manifolds. So, just as one can find the value of an integral $(f dx = dF)$ over a 1-dimensional manifold $([a, b])$ by considering the anti-derivative (F) at the 0-dimensional boundaries $(\{a, b\})$, one can generalize the fundamental theorem of calculus, with a few additional caveats, to deal with the value of integrals $(d\omega)$ over n-dimensional manifolds (Ω) by considering the antiderivative (ω) at the $(n - 1)$-dimensional boundaries $(\partial\Omega)$ of the manifold.

So the fundamental theorem reads:

$$\int_{[a,b]} f(x)dx = \int_{[a,b]} dF = \int_{\{a\}^- \cup \{b\}^+} F = F(b) - F(a).$$

Formulation for Smooth Manifolds with Boundary

Let Ω be an oriented smooth manifold with boundary of dimension n and let α be a smooth n-differential form that is compactly supported on Ω. First, suppose that α is compactly supported in the domain of a single, oriented coordinate chart $\{U, \varphi\}$. In this case, we define the integral of α over Ω as,

$$\int_\Omega \alpha = \int_{\varphi(U)} \left(\varphi^{-1}\right)^* \alpha,$$

i.e., via the pullback of α to R^n.

More generally, the integral of α over Ω is defined as follows: Let $\{\psi_i\}$ be a partition of unity associated with a locally finite cover $\{U_i, \varphi_i\}$ of (consistently oriented) coordinate charts, then define the integral

$$\int_\Omega \alpha \equiv \sum_i \int_{U_i} \psi_i \alpha,$$

where each term in the sum is evaluated by pulling back to R^n as described above. This quantity is well-defined; that is, it does not depend on the choice of the coordinate charts, nor the partition of unity.

Theorem: (Stokes–Cartan) If ω is a smooth $(n-1)$-form with compact support on smooth n-dimensional manifold-with-boundary Ω, $\partial\Omega$ denotes the boundary of Ω given the induced orientation, and $i : \partial\Omega \to \Omega$ is the inclusion map, then:

$$\int_\Omega d\omega = \int_{\partial\Omega} i^* \omega \, .$$

Conventionally, $\int_{\partial\Omega} i^* \omega$ is abbreviated as $\int_{\partial\Omega} \omega$, since the pullback of a differential form by the inclusion map is simply its restriction to its domain: $i^* \omega = \omega|_{\partial\Omega}$. Here d is the exterior derivative, which is defined using the manifold structure only. The right-hand side is sometimes written as $\oint_{\partial\Omega} \omega$ to stress the fact that the $(n-1)$-manifold $\partial\Omega$ has no boundary. (This fact is also an implication of Stokes' theorem, since for a given smooth n-dimensional manifold Ω, application of the theorem twice gives $\int_{\partial(\partial\Omega)} \omega = \int_\Omega d(d\omega) = 0$ for any $(n-2)$-form ω, which implies that $\partial(\partial\Omega) = \emptyset$.) The right-hand side of the equation is often used to formulate *integral* laws; the left-hand side then leads to equivalent *differential* formulations.

The theorem is often used in situations where Ω is an embedded oriented submanifold of some bigger manifold, often R^k, on which the form ω is defined.

Topological Preliminaries: Integration over Chains

Let M be a smooth manifold. A (smooth) singular k-simplex in M is defined as a smooth map from the standard simplex in R^k to M. The group $C_k(M, Z)$ of singular k-chains on M is defined to be the free abelian group on the set of singular k-simplices in M. These groups, together with the boundary map, ∂, define a chain complex. The corresponding homology (resp. cohomology) group is isomorphic to the usual singular homology group $H_k(M, Z)$ (resp. the singular cohomology group $H^k(M, Z)$), defined using continuous rather than smooth simplices in M.

On the other hand, the differential forms, with exterior derivative, d, as the connecting map, form a cochain complex, which defines the de Rham cohomology groups $H^k_{dR}(M, R)$.

Differential k-forms can be integrated over a k-simplex in a natural way, by pulling back to R^k. Extending by linearity allows one to integrate over chains. This gives a linear map from the space of k-forms to the kth group of singular cochains, $C^k(M, Z)$, the linear functionals on $C_k(M, Z)$. In other words, a k-form ω defines a functional:

$$I(\omega)(c) = \oint_c \omega$$

on the k-chains. Stokes' theorem says that this is a chain map from de Rham cohomology to singular cohomology with real coefficients; the exterior derivative, d, behaves like the *dual* of ∂ on forms. This gives a homomorphism from de Rham cohomology to singular cohomology. On the level of forms, this means:

- Closed forms, i.e., $d\omega = 0$, have zero integral over *boundaries*, i.e. over manifolds that can be written as $\partial\sum_c M_c$,

- Exact forms, i.e., $\omega = d\sigma$, have zero integral over *cycles*, i.e. if the boundaries sum up to the empty set: $\sum_c M_c = \emptyset$.

De Rham's theorem shows that this homomorphism is in fact an isomorphism. So the converse to points 1 and 2 above hold true. In other words, if $\{c_i\}$ are cycles generating the kth homology group, then for any corresponding real numbers, $\{a_i\}$, there exist a closed form, ω, such that:

$$\oint_{c_i} \omega = a_i,$$

and this form is unique up to exact forms.

Stokes' theorem on smooth manifolds can be derived from Stokes' theorem for chains in smooth manifolds, and vice versa.

Theorem: (Stokes' theorem for chains) If c is a smooth k-chain in a smooth manifold M, and ω is a smooth $(k-1)$-form on M, then:

$$\int_{\partial c} \omega = \int_c d\omega \cdot$$

Underlying Principle

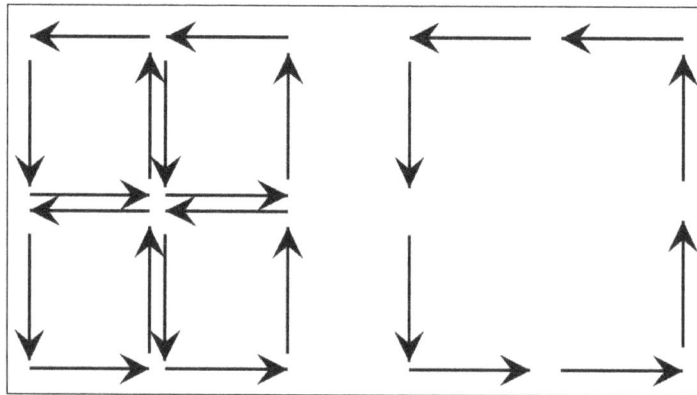

To simplify these topological arguments, it is worthwhile to examine the underlying principle by considering an example for $d = 2$ dimensions. The essential idea can be understood by the diagram, in an oriented tiling of a manifold, the interior paths are traversed in opposite directions; their contributions to the path integral thus cancel each other pairwise. As a consequence, only the contribution from the boundary remains. It thus suffices to prove Stokes' theorem for sufficiently fine tilings (or, equivalently, simplices), which usually is not difficult.

DIVERGENCE THEOREM

The divergence theorem, more commonly known as Gauss's theorem and also known as the Gauss-Ostrogradsky theorem, is a theorem in vector calculus that can be stated as follows. Let V be a region in space with boundary ∂v. Then the volume integral of the divergence $\nabla \cdot F$ of F over v and the surface integral of F over the boundary ∂v of V are related by,

$$\int_v (\nabla \cdot F) dV = \int_{\partial v} F \cdot d\,a.$$

The divergence theorem is a mathematical statement of the physical fact that, in the absence of the creation or destruction of matter, the density within a region of space can change only by having it flow into or away from the region through its boundary.

A special case of the divergence theorem follows by specializing to the plane. Letting S be a region in the plane with boundary ∂S, equation $\int_v (\nabla \cdot F) dV = \int_{av} F \cdot da$. then collapses to,

$$\int_s \nabla \cdot F\, dA = \int_{as} F \cdot \hat{n}\, ds.$$

If the vector field F satisfies certain constraints, simplified forms can be used. For example, if $F(x, y, z) = v(x, y, z)c$ where c is a constant vector $\neq 0$, then,

$$\int_s F \cdot da = c \cdot \int_s v\, da.$$

But,

$$\nabla \cdot (f\, v) = (\nabla f) \cdot v + f (\nabla . v),$$

so,

$$\int_v \nabla \cdot (cv) dV = \int_v [(\nabla v) \cdot c + v \nabla \cdot c] dV$$
$$= c \cdot \int_v \nabla v\, dV$$

and

$$c \cdot \left(\int_s v\, da - \int_v \nabla v\, dV \right) = 0$$

But $c \neq 0$, and $c \cdot f(v)$ must vary with v so that $c \cdot f(v)$ cannot always equal zero. Therefore,

$$\int_s v\, da = \int_v \nabla v\, dV.$$

Similarly, if $F(x, y, z) = c \times P(x, y, z)$, where c is a constant vector $\neq 0$, then,

$$\int_s da \times P = \int_v \nabla \times P\, dV.$$

References

- Vector-calculus, boundless-calculus: courses.lumenlearning.com, Retrieved 19 July, 2019
- Linear-Algebra-Gradient-of-a-Scalar-Field, Math-Formulas: web-formulas.com, Retrieved 15 April, 2019
- Divergence, multivariable-calculus-multivariable-derivatives-divergence-and-curl-articles: khanacademy.org, Retrieved 09 August, 2019
- Hazewinkel, Michiel (2001). Encyclopedia of Mathematics. Springer. pp. Surface Integral. ISBN 978-1-55608-010-4
- Curl: mathworld.wolfram.com, Retrieved 28 January, 2019
- DivergenceTheorem: mathworld.wolfram.com, Retrieved 19 June, 2019

Differential Equations

Differential equations are the mathematical equations that relate some functions with their derivatives. They are divided into ordinary differential equations and partial differential equations. The topics elaborated in this chapter will help in gaining a better perspective about these types of differential equation.

A differential equation is any equation which contains derivatives, either ordinary derivatives or partial derivatives.

There is one differential equation that everybody probably knows, that is Newton's Second Law of Motion. If an object of mass m is moving with acceleration a and being acted on with force F then Newton's Second Law tells us.

$$F = ma$$

To see that this is in fact a differential equation we need to rewrite it a little. First, remember that we can rewrite the acceleration, a, in one of two ways.

$$a = \frac{dv}{dt} \quad \text{OR} \quad a = \frac{d^2u}{dt^2}$$

Where v is the velocity of the object and uu is the position function of the object at any time t. We should also remember at this point that the force, F may also be a function of time, velocity, and/or position.

So, with all these things in mind Newton's Second Law can now be written as a differential equation in terms of either the velocity, v, or the position, u, of the object as follows.

$$m\frac{dv}{dt} = F(t,v)$$

$$m\frac{d^2u}{dt^2} = F\left(t,u,\frac{du}{dt}\right)$$

So, here is our first differential equation.

Here are a few more examples of differential equations.

$$ay'' + by' + cy = g(t)$$

$$\sin(y)\frac{d^2y}{dx^2} = (1-y)\frac{dy}{dx} + y^2 e^{-5y}$$

$$y^{(4)} + 10y''' - 4y' + 2y = \cos(t)$$

$$\alpha^2 \frac{\partial^2 u}{\partial x^2} = \frac{\partial u}{\partial t}$$

$$a^2 u_{xx} = u_{tt}$$

$$\frac{\partial^3 u}{\partial^2 x \partial t} = 1 + \frac{\partial u}{\partial y}$$

Order

The order of a differential equation is the largest derivative present in the differential equation. In the differential equations listed above $m\dfrac{dv}{dt} = F(t,v)$ is a first order differential equation,

$m\dfrac{d^2 u}{dt^2} = F\left(t, u, \dfrac{du}{dt}\right)$, $ay'' + by' + cy = g(t)$, $\sin(y)\dfrac{d^2 y}{dx^2} = (1-y)\dfrac{dy}{dx} + y^2 e^{-5y}$, $\alpha^2 \dfrac{\partial^2 u}{\partial x^2} = \dfrac{\partial u}{\partial t}$, and

$a^2 u_{xx} = u_{tt}$ are second order differential equations, $\dfrac{\partial^3 u}{\partial^2 x \partial t} = 1 + \dfrac{\partial u}{\partial y}$ is a third order differential

equation and $y^{(4)} + 10y''' - 4y' + 2y = \cos(t)$ is a fourth order differential equation.

Note that the order does not depend on whether or not you've got ordinary or partial derivatives in the differential equation.

We will be looking almost exclusively at first and second order differential equations in these notes. As you will see most of the solution techniques for second order differential equations can be easily (and naturally) extended to higher order differential equations.

Ordinary and Partial Differential Equations

A differential equation is called an ordinary differential equation, abbreviated by ode, if it has ordinary derivatives in it. Likewise, a differential equation is called a partial differential equation, abbreviated by pde, if it has partial derivatives in it.

Linear Differential Equations

A linear differential equation is any differential equation that can be written in the following form.

$$a_n(t)y^{(n)}(t) + a_{n-1}(t)y^{(n-1)}(t) + \cdots + a_1(t)y'(t) + a_0(t)y(t) = g(t)$$

The important thing to note about linear differential equations is that there are no products of the function, $y(t)$, and its derivatives and neither the function or its derivatives occur to any power other than the first power. Also note that neither the function or its derivatives are "inside" another function, for example, $\sqrt{y'}$ or e^y.

The coefficients $a_0(t), \ldots, a_n(t)$ and $g(t)$ can be zero or non-zero functions, constant or non-constant functions, linear or non-linear functions. Only the function, $y(t)$ and its derivatives are used in determining if a differential equation is linear.

If a differential equation cannot be written in the form, $a_n(t)y^{(n)}(t)+a_{n-1}(t)y^{(n-1)}(t)$ $+\cdots+a_1(t)y'(t)+a_0(t)y(t)=g(t)$ then it is called a non-linear differential eqution.

A solution to a differential equation on an interval $\alpha<t<\beta$ is any functiony (t) which satisfies the differential equation in question on the interval $\alpha<t<\beta$. It is important to note that solutions are often accompanied by intervals and these intervals can impart some important information about the solution.

So, even though a function may symbolically satisfy a differential equation, because of certain restrictions brought about by the solution we cannot use all values of the independent variable and hence, must make a restriction on the independent variable. This will be the case with many solutions to differential equations.

Note that there are in fact many more possible solutions to the differential equation given. For instance, all of the following are also solutions,

$$y(x) = x^{-\frac{1}{2}}$$

$$y(x) = -9x^{-\frac{3}{2}}$$

$$y(x) = 7x^{-\frac{1}{2}}$$

$$y(x) = -9x^{-\frac{3}{2}} + 7x^{-\frac{1}{2}}$$

We'll leave the details to you to check that these are in fact solutions. Given these examples can you come up with any other solutions to the differential equation?

So, given that there are an infinite number of solutions to the differential equation in the last example we can ask a natural question. Which is the solution that we want or does it matter which solution we use? This question leads us to the next definition in this part.

Initial Condition

Initial Condition(s) are a condition, or set of conditions, on the solution that will allow us to determine which solution that we are after. Initial conditions are of the form,

$$y(t_0) = y_0 \text{ and} / \text{or } y^{(k)}(t_0) = y_k$$

So, in other words, initial conditions are values of the solution and/or its derivative(s) at specific points. As we will see eventually, solutions to "nice enough" differential equations are unique and hence only one solution will meet the given initial conditions.

The number of initial conditions that are required for a given differential equation will depend upon the order of the differential equation as we will see.

Initial Value Problem

An Initial Value Problem (or IVP) is a differential equation along with an appropriate number of initial conditions.

As we noted earlier the number of initial conditions required will depend on the order of the differential equation.

Interval of Validity

The interval of validity for an IVP with initial condition(s):

$$y(t_0) = y_0 \text{ and/or } y^{(k)}(t_0) = y_k$$

Is the largest possible interval on which the solution is valid and contains t_0. These are easy to define, but can be difficult to find, so we're going to put off saying anything more about these until we get into actually solving differential equations and need the interval of validity.

General Solution

The general solution to a differential equation is the most general form that the solution can take and doesn't take any initial conditions into account.

Actual Solution

The actual solution to a differential equation is the specific solution that not only satisfies the differential equation, but also satisfies the given initial condition(s).

From this last example we can see that once we have the general solution to a differential equation finding the actual solution is nothing more than applying the initial condition(s) and solving for the constant(s) that are in the general solution.

Implicit/Explicit Solution

In this case it's easier to define an explicit solution, then tell you what an implicit solution isn't, and then give you an example to show you the difference. So, that's what we'll do.

An explicit solution is any solution that is given in the form y=y (t). In other words, the only place that y actually shows up is once on the left side and only raised to the first power. An implicit solution is any solution that isn't in explicit form. Note that it is possible to have either general implicit/explicit solutions and actual implicit/explicit solutions.

In this case we were able to find an explicit solution to the differential equation. It should be noted however that it will not always be possible to find an explicit solution.

Also, note that in this case we were only able to get the explicit actual solution because we had the initial condition to help us determine which of the two functions would be the correct solution.

ORDINARY DIFFERENTIAL EQUATIONS

In mathematics, an ordinary differential equation (ODE) is a differential equation containing one or more functions of one independent variable and the derivatives of those functions. The term ordinary is used in contrast with the term partial differential equation which may be with respect to more than one independent variable.

Differential Equations

A linear differential equation is a differential equation that is defined by a linear polynomial in the unknown function and its derivatives, that is an equation of the form:

$$a_0(x)y + a_1(x)y' + a_2(x)y'' + \cdots + a_n(x)y^{(n)} + b(x) = 0,$$

where $a_0(x), \ldots, a_n(x)$ and $b(x)$ are arbitrary differentiable functions that do not need to be linear, and $y', \ldots, y^{(n)}$ are the successive derivatives of the unknown function y of the variable x.

Among ordinary differential equations, linear differential equations play a prominent role for several reasons. Most elementary and special functions that are encountered in physics and applied mathematics are solutions of linear differential equations. When physical phenomena are modeled with non-linear equations, they are generally approximated by linear differential equations for an easier solution. The few non-linear ODEs that can be solved explicitly are generally solved by transforming the equation into an equivalent linear ODE.

Some ODEs can be solved explicitly in terms of known functions and integrals. When that is not possible, the equation for computing the Taylor series of the solutions may be useful. For applied problems, numerical methods for ordinary differential equations can supply an approximation of the solution.

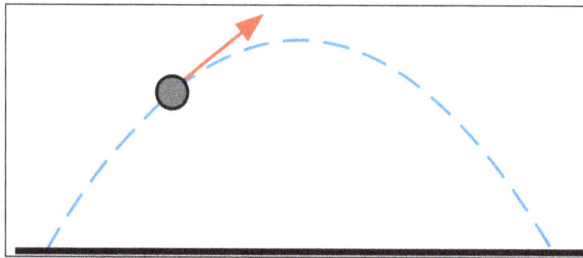

The trajectory of a projectile launched from a cannon follows a curve determined by an ordinary differential equation that is derived from Newton's second law.

Ordinary differential equations (ODEs) arise in many contexts of mathematics and social and natural sciences. Mathematical descriptions of change use differentials and derivatives. Various differentials, derivatives, and functions become related via equations, such that a differential equation is a result that describes dynamically changing phenomena, evolution, and variation. Often, quantities are defined as the rate of change of other quantities (for example, derivatives of displacement with respect to time), or gradients of quantities, which is how they enter differential equations.

Specific mathematical fields include geometry and analytical mechanics. Scientific fields include much of physics and astronomy (celestial mechanics), meteorology (weather modelling), chemistry (reaction rates), biology (infectious diseases, genetic variation), ecology and population modelling (population competition), economics (stock trends, interest rates and the market equilibrium price changes).

Many mathematicians have studied differential equations and contributed to the field, including Newton, Leibniz, the Bernoulli family, Riccati, Clairaut, d'Alembert, and Euler.

A simple example is Newton's second law of motion — the relationship between the displacement x and the time t of an object under the force F, is given by the differential equation:

$$m\frac{d^2x(t)}{dt^2} = F(x(t))$$

which constrains the motion of a particle of constant mass m. In general, F is a function of the position x(t) of the particle at time t. The unknown function x(t) appears on both sides of the differential equation, and is indicated in the notation F(x(t)).

In what follows, let y be a dependent variable and x an independent variable, and y = f(x) is an unknown function of x. The notation for differentiation varies depending upon the author and upon which notation is most useful for the task at hand. In this context, the Leibniz's notation $(dy/dx, d^2y/dx^2, ..., d^ny/dx^n)$ is more useful for differentiation and integration, whereas Lagrange's notation $(y', y'', ..., y^{(n)})$ is more useful for representing derivatives of any order compactly, and Newton's notation $(\dot{y}, \ddot{y}, \dddot{y})$ is often used in physics for representing derivatives of low order with respect to time.

Given F, a function of x, y, and derivatives of y. Then an equation of the form:

$$F\left(x, y, y', ..., y^{(n-1)}\right) = y^{(n)}$$

is called an explicit ordinary differential equation of order n.

More generally, an implicit ordinary differential equation of order n takes the form:

$$F\left(x, y, y', y'', ..., y^{(n)}\right) = 0$$

There are further classifications:

Autonomous:

- A differential equation not depending on x is called autonomous.

Linear:

- A differential equation is said to be linear if F can be written as a linear combination of the derivatives of y:

$$y^{(n)} = \sum_{i=0}^{n-1} a_i(x)y^{(i)} + r(x)$$

where ai(x) and r(x) are continuous functions in x. The function r(x) is called the source term.

Homogeneous:

- If r(x) = 0, and consequently one "automatic" solution is the trivial solution, y = 0. The solution of a linear homogeneous equation is a complementary function, denoted here by y_c.

Nonhomogeneous (or inhomogeneous):

- If r(x) ≠ 0. The additional solution to the complementary function is the particular integral, denoted here by y_p.

The general solution to a linear equation can be written as $y = y_c + y_p$.

Non-linear:

- A differential equation that cannot be written in the form of a linear combination.

System of ODEs

A number of coupled differential equations form a system of equations. If **y** is a vector whose elements are functions; $\mathbf{y}(x) = [y_1(x), y_2(x),..., y_m(x)]$, and F is a vector-valued function of **y** and its derivatives, then:

$$\mathbf{y}^{(n)} = \mathbf{F}\left(x,\mathbf{y},\mathbf{y}',\mathbf{y}'',...,\mathbf{y}^{(n-1)}\right)$$

is an explicit system of ordinary differential equations of order n and dimension m. In column vector form:

$$\begin{pmatrix} y_1^{(n)} \\ y_2^{(n)} \\ \vdots \\ y_m^{(n)} \end{pmatrix} = \begin{pmatrix} f_1\left(x,\mathbf{y},\mathbf{y}',\mathbf{y}'',...,\mathbf{y}^{(n-1)}\right) \\ f_2\left(x,\mathbf{y},\mathbf{y}',\mathbf{y}'',...,\mathbf{y}^{(n-1)}\right) \\ \vdots \\ f_m\left(x,\mathbf{y},\mathbf{y}',\mathbf{y}'',...,\mathbf{y}^{(n-1)}\right) \end{pmatrix}$$

These are not necessarily linear. The implicit analogue is:

$$\mathbf{F}\left(x,\mathbf{y},\mathbf{y}',\mathbf{y}'',...,\mathbf{y}^{(n)}\right) = \mathbf{0}$$

where 0 = (0, 0,... , 0) is the zero vector. In matrix form:

$$\begin{pmatrix} f_1(x,\mathbf{y},\mathbf{y}',\mathbf{y}'',...,\mathbf{y}^{(n)}) \\ f_2(x,\mathbf{y},\mathbf{y}',\mathbf{y}'',...,\mathbf{y}^{(n)}) \\ \vdots \\ f_m(x,\mathbf{y},\mathbf{y}',\mathbf{y}'',...,\mathbf{y}^{(n)}) \end{pmatrix} = \begin{pmatrix} 0 \\ 0 \\ \vdots \\ 0 \end{pmatrix}$$

For a system of the form $\mathbf{F}(x, \mathbf{y}, \mathbf{y}') = \mathbf{0}$, some sources also require that the Jacobian matrix $\dfrac{\partial \mathbf{F}(x, \mathbf{u}, \mathbf{v})}{\partial \mathbf{v}}$ be non-singular in order to call this an implicit ODE [system]; an implicit ODE system satisfying this Jacobian non-singularity condition can be transformed into an explicit ODE system. In the same sources, implicit ODE systems with a singular Jacobian are termed differential algebraic equations (DAEs). This distinction is not merely one of terminology; DAEs have fundamentally different characteristics and are generally more involved to solve than (nonsingular) ODE systems. Presumably for additional derivatives, the Hessian matrix and so forth are also assumed non-singular according to this scheme, although note that any ODE of order greater than one can be [and usually is] rewritten as system of ODEs of first order, which makes the Jacobian singularity criterion sufficient for this taxonomy to be comprehensive at all orders.

The behavior of a system of ODEs can be visualized through the use of a phase portrait.

Given a differential equation,

$$F\left(x, y, y', \ldots, y^{(n)}\right) = 0$$

a function u: I ⊂ R → R is called a solution or integral curve for F, if u is n-times differentiable on I, and,

$$F(x, u, u', \ldots, u^{(n)}) = 0 \quad x \in I.$$

Given two solutions u: J ⊂ R → R and v: I ⊂ R → R, u is called an extension of v if I ⊂ J and,

$$u(x) = v(x) \quad x \in I.$$

A solution that has no extension is called a maximal solution. A solution defined on all of R is called a global solution.

A general solution of an nth-order equation is a solution containing n arbitrary independent constants of integration. A particular solution is derived from the general solution by setting the constants to particular values, often chosen to fulfill set 'initial conditions or boundary conditions'. A singular solution is a solution that cannot be obtained by assigning definite values to the arbitrary constants in the general solution.

Theories

Singular Solutions

The theory of singular solutions of ordinary and partial differential equations was a subject of research from the time of Leibniz, but only since the middle of the nineteenth century has it received special attention. A valuable but little-known work on the subject is that of Houtain. Darboux was a leader in the theory, and in the geometric interpretation of these solutions he opened a field worked by various writers, notably Casorati and Cayley. To the latter is due the theory of singular solutions of differential equations of the first order as accepted circa 1900.

Reduction to Quadratures

The primitive attempt in dealing with differential equations had in view a reduction to quadratures. As it had been the hope of eighteenth-century algebraists to find a method for solving the general equation of the nth degree, so it was the hope of analysts to find a general method for integrating any differential equation. Gauss showed, however, that complex differential equations require complex numbers. Hence, analysts began to substitute the study of functions, thus opening a new and fertile field. Cauchy was the first to appreciate the importance of this view. Thereafter, the real question was no longer whether a solution is possible by means of known functions or their integrals, but whether a given differential equation suffices for the definition of a function of the independent variable or variables, and, if so, what are the characteristic properties.

Fuchsian Theory

Two memoirs by Fuchs inspired a novel approach, subsequently elaborated by Thomé and Frobenius. Collet was a prominent contributor beginning in 1869. His method for integrating a non-linear system was communicated to Bertrand in 1868. Clebsch attacked the theory along lines parallel to those in his theory of Abelian integrals. As the latter can be classified according to the properties of the fundamental curve that remains unchanged under a rational transformation, Clebsch proposed to classify the transcendent functions defined by differential equations according to the invariant properties of the corresponding surfaces f = 0 under rational one-to-one transformations.

Lie's Theory

From 1870, Sophus Lie's work put the theory of differential equations on a better foundation. He showed that the integration theories of the older mathematicians can, using Lie groups, be referred to a common source, and that ordinary differential equations that admit the same infinitesimal transformations present comparable integration difficulties. He also emphasized the subject of transformations of contact.

Lie's group theory of differential equations has been certified, namely: (1) that it unifies the many ad hoc methods known for solving differential equations, and (2) that it provides powerful new ways to find solutions. The theory has applications to both ordinary and partial differential equations.

A general solution approach uses the symmetry property of differential equations, the continuous infinitesimal transformations of solutions to solutions (Lie theory). Continuous group theory, Lie algebras, and differential geometry are used to understand the structure of linear and nonlinear (partial) differential equations for generating integrable equations, to find its Lax pairs, recursion operators, Bäcklund transform, and finally finding exact analytic solutions to DE.

Symmetry methods have been applied to differential equations that arise in mathematics, physics, engineering, and other disciplines.

Sturm–Liouville Theory

Sturm–Liouville theory is a theory of a special type of second order linear ordinary differential equation. Their solutions are based on eigenvalues and corresponding eigenfunctions of linear operators defined via second-order homogeneous linear equations. The problems are identified as

Sturm-Liouville Problems (SLP) and are named after J.C.F. Sturm and J. Liouville, who studied them in the mid-1800s. SLPs have an infinite number of eigenvalues, and the corresponding eigenfunctions form a complete, orthogonal set, which makes orthogonal expansions possible. This is a key idea in applied mathematics, physics, and engineering. SLPs are also useful in the analysis of certain partial differential equations.

Existence and Uniqueness of Solutions

There are several theorems that establish existence and uniqueness of solutions to initial value problems involving ODEs both locally and globally. The two main theorems are:

In their basic form both of these theorems only guarantee local results, though the latter can be extended to give a global result, for example, if the conditions of Grönwall's inequality are met.

Also, uniqueness theorems like the Lipschitz one above do not apply to DAE systems, which may have multiple solutions stemming from their (non-linear) algebraic part alone.

Theorem	Assumption	Conclusion
Peano existence theorem	F continuous	local existence only
Picard–Lindelöf theorem	F Lipschitz continuous	local existence and uniqueness

Local Existence and Uniqueness Theorem Simplified

The theorem can be stated simply as follows. For the equation and initial value problem:

$$y' = F(x, y), \quad y_0 = y(x_0)$$

if F and $\partial F/\partial y$ are continuous in a closed rectangle:

$$R = [x_0 - a, x_0 + a] \times [y_0 - b, y_0 + b]$$

in the x-y plane, where a and b are real (symbolically: a, b ∈ R) and × denotes the cartesian product, square brackets denote closed intervals, then there is an interval

$$I = [x_0 - h, x_0 + h] \subset [x_0 - a, x_0 + a]$$

for some h ∈ R where the solution to the above equation and initial value problem can be found. That is, there is a solution and it is unique. Since there is no restriction on F to be linear, this applies to non-linear equations that take the form F(x, y), and it can also be applied to systems of equations.

Global Uniqueness and Maximum Domain of Solution

When the hypotheses of the Picard–Lindelöf theorem are satisfied, then local existence and uniqueness can be extended to a global result.

For each initial condition (x_0, y_0) there exists a unique maximum (possibly infinite) open interval

$$I_{max} = (x_-, x_+), x_\pm \in \mathbb{R} \cup \{\pm\infty\}, x_0 \in I_{max}$$

such that any solution that satisfies this initial condition is a restriction of the solution that satisfies this initial condition with domain I_{max} .

In the case that $x_{\pm} \neq \pm\infty$, there are exactly two possibilities:

- Explosion in finite time: $\limsup\limits_{x \to x_{\pm}} \| y(x) \| \to \infty$

- Leaves domain of definition: $\lim\limits_{x \to x_{\pm}} y(x) \in \partial\overline{\Omega}$

where Ω is the open set in which F is defined, and $\partial\overline{\Omega}$ is its boundary.

Note that the maximum domain of the solution:

- Is always an interval (to have uniqueness)

- May be smaller than $\mathbb{R}$

- May depend on the specific choice of (x_0, y_0)

Example:

$$y' = y^2$$

This means that F(x, y) = y², which is C¹ and therefore locally Lipschitz continuous, satisfying the Picard–Lindelöf theorem.

Even in such a simple setting, the maximum domain of solution cannot be all $\mathbb{R}$ since the solution is

$$y(x) = \frac{y_0}{(x_0 - x)y_0 + 1}$$

which has maximum domain:

$$\begin{cases} \mathbb{R} & y_0 = 0 \\ \left(-\infty, x_0 + \dfrac{1}{y_0} \right) & y_0 > 0 \\ \left(x_0 + \dfrac{1}{y_0}, +\infty \right) & y_0 < 0 \end{cases}$$

This shows clearly that the maximum interval may depend on the initial conditions. The domain of y could be taken as being $\mathbb{R} \setminus (x_0 + 1 / y_0)$, but this would lead to a domain that is not an interval, so that the side opposite to the initial condition would be disconnected from the initial condition, and therefore not uniquely determined by it.

The maximum domain is not $\mathbb{R}$ because:

$$\lim\limits_{x \to x_{\pm}} \| y(x) \| \to \infty,$$

which is one of the two possible cases according to the above theorem.

Reduction of Order

Differential equations can usually be solved more easily if the order of the equation can be reduced.

Reduction to a First-order System

Any explicit differential equation of order n,

$$F\left(x, y, y', y'', \ldots, y^{(n-1)}\right) = y^{(n)}$$

can be written as a system of n first-order differential equations by defining a new family of unknown functions:

$$y_i = y^{(i-1)}.$$

for i = 1, 2,..., n. The n-dimensional system of first-order coupled differential equations is then,

$$
\begin{aligned}
y_1' &= y_2 \\
y_2' &= y_3 \\
&\vdots \\
y_{n-1}' &= y_n \\
y_n' &= F(x, y_1, \ldots, y_n).
\end{aligned}
$$

more compactly in vector notation:

$$\mathbf{y}' = \mathbf{F}(x, \mathbf{y})$$

where,

$$\mathbf{y} = (y_1, \ldots, y_n), \quad \mathbf{F}(x, y_1, \ldots, y_n) = (y_2, \ldots, y_n, F(x, y_1, \ldots, y_n)).$$

Some differential equations have solutions that can be written in an exact and closed form. Several important classes are given here.

In the table below, P(x), Q(x), P(y), Q(y), and M(x,y), N(x,y) are any integrable functions of x, y, and b and c are real given constants, and C_1, C_2,... are arbitrary constants (complex in general). The differential equations are in their equivalent and alternative forms that lead to the solution through integration.

In the integral solutions, λ and ε are dummy variables of integration (the continuum analogues of indices in summation), and the notation $\int^x F(\lambda) \, d\lambda$ just means to integrate $F(\lambda)$ with respect to λ, then after the integration substitute $\lambda = x$, without adding constants (explicitly stated).

Type	Differential equation	Solution method	General solution
Separable	First-order, separable in x and y (general case,) $P_1(x)Q_1(y)+P_2(x)Q_2(y)\dfrac{dy}{dx}=0$ $P_1(x)Q_1(y)dx+P_2(x)Q_2(y)dy=0$	Separation of variables (divide by P_2Q_1).	$\displaystyle\int^x \frac{P_1(\lambda)}{P_2(\lambda)}d\lambda+\int^y \frac{Q_2(\lambda)}{Q_1(\lambda)}d\lambda=C$
	First-order, separable in x $\dfrac{dy}{dx}=F(x)$ $dy=F(x)dx$	Direct integration.	$\displaystyle y=\int^x F(\lambda)d\lambda+C$
	First-order, autonomous, separable in y $\dfrac{dy}{dx}=F(y)$ $dy=F(y)dx$	Separation of variables (divide by F).	$\displaystyle x=\int^y \frac{d\lambda}{F(\lambda)}+C$
	First-order, separable in x and y $P(y)\dfrac{dy}{dx}+Q(x)=0$ $P(y)dy+Q(x)dx=0$	Integrate throughout.	$\displaystyle\int^y P(\lambda)d\lambda+\int^x Q(\lambda)d\lambda=C$
General first-order	First-order, homogeneous $\dfrac{dy}{dx}=F\left(\dfrac{y}{x}\right)$	Set y = ux, then solve by separation of variables in u and x.	$\displaystyle\ln(Cx)=\int^{y/x} \frac{d\lambda}{F(\lambda)-\lambda}$
	First-order, separable $yM(xy)+xN(xy)\dfrac{dy}{dx}=0$ $yM(xy)dx+xN(xy)dy=0$	Separation of variables (divide by xy).	$\displaystyle\ln(Cx)=\int^{xy} \frac{N(\lambda)d\lambda}{\lambda[N(\lambda)-M(\lambda)]}$ If N = M, the solution is xy = C.
	Exact differential, first-order $M(x,y)\dfrac{dy}{dx}+N(x,y)=0$ $M(x,y)dy+N(x,y)dx=0$ where $\dfrac{\partial M}{\partial x}=\dfrac{\partial N}{\partial y}$	Integrate throughout.	$\displaystyle F(x,y)=\int^y M(x,\lambda)d\lambda+\int^x N(\lambda,y)d\lambda$ $+Y(y)+X(x)=C$ where Y(y) and X(x) are functions from the integrals rather than constant values, which are set to make the final function F(x, y) satisfy the initial equation.
	Inexact differential, first-order $M(x,y)\dfrac{dy}{dx}+N(x,y)=0$ $M(x,y)dy+N(x,y)dx=0$ where $\dfrac{\partial M}{\partial x}\neq\dfrac{\partial N}{\partial y}$	Integration factor $\mu(x,y)$ satisfying $\dfrac{\partial(\mu M)}{\partial x}=\dfrac{\partial(\mu N)}{\partial y}$	If $\mu(x,y)$ can be found: $\displaystyle F(x,y)=\int^y \mu(x,\lambda)M(x,\lambda)d\lambda+$ $\displaystyle\int^x \mu(\lambda,y)N(\lambda,y)d\lambda$ $+Y(y)+X(x)=C$

General second-order	Second-order, autonomous $$\frac{d^2y}{dx^2}=F(y)$$	Multiply both sides of equation by 2dy/dx, substitute $2\dfrac{dy}{dx}\dfrac{d^2y}{dx^2}=\dfrac{d}{dx}\left(\dfrac{dy}{dx}\right)^2$, then integrate twice.	$$x=\pm\int^y\frac{d\lambda}{\sqrt{2\int^\lambda F(\varepsilon)d\varepsilon+C_1}}+C_2$$
Linear to nth order	First-order, linear, inhomogeneous, function coefficients $$\frac{dy}{dx}+P(x)y=Q(x)$$	Integrating factor: $$e^{\int^x P(\lambda)d\lambda}$$	$$y=e^{-\int^x P(\lambda)d\lambda}\left[\int^x e^{\int^\lambda P(\varepsilon)d\varepsilon}Q(\lambda)d\lambda+C\right]$$
	Second-order, linear, inhomogeneous, constant coefficients $$\frac{d^2y}{dx^2}+b\frac{dy}{dx}+cy=r(x)$$	Complementary function y_c: assume $y_c=e^{\alpha x}$, substitute and solve polynomial in α, to find the linearly independent functions $e^{\alpha_j x}$. Particular integral y_p: in general the method of variation of parameters, though for very simple r(x) inspection may work.	$y=y_c+y_p$ If b² > 4c, then $$y_c=C_1 e^{-\frac{x}{2}\left(b+\sqrt{b^2-4c}\right)}+C_2 e^{-\frac{x}{2}\left(b-\sqrt{b^2-4c}\right)}$$ If b² = 4c, then $$y_c=(C_1 x+C_2)e^{-\frac{bx}{2}}$$ If b² < 4c, then $$y_c=e^{-\frac{bx}{2}}\left[C_1\sin\left(x\frac{\sqrt{4c-b^2}}{2}\right)+C_2\cos\left(x\frac{\sqrt{4c-b^2}}{2}\right)\right]$$
	nth-order, linear, inhomogeneous, constant coefficients $$\sum_{j=0}^n b_j\frac{d^j y}{dx^j}=r(x)$$	Complementary function y_c: assume $y_c=e^{\alpha x}$, substitute and solve polynomial in α, to find the linearly independent functions $e^{\alpha_j x}$. Particular integral y_p: in general the method of variation of parameters, though for very simple r(x) inspection may work.	$y=y_c+y_p$ Since α_j are the solutions of the polynomial of degree n: $\prod_{j=1}^n(\alpha-\alpha_j)=0$, then: for α_j all different, $$y_c=\sum_{j=1}^n C_j e^{\alpha_j x}$$ for each root α_j repeated k_j times, $$y_c=\sum_{j=1}^n\left(\sum_{\ell=1}^{k_j}C_{j,\ell}x^{\ell-1}\right)e^{\alpha_j x}$$ for some α_j complex, then setting $\alpha=\chi_j+i\gamma_j$, and using Euler's formula, allows some terms in the previous results to be written in the form $$C_j e^{\alpha_j x}=C_j e^{\chi_j x}\cos(\gamma_j x+\varphi_j)$$ where ϕ_j is an arbitrary constant (phase shift).

Series Solutions

Power Series Solution of Differential Equations

In mathematics, the power series method is used to seek a power series solution to certain differential equations. In general, such a solution assumes a power series with unknown coefficients, then substitutes that solution into the differential equation to find a recurrence relation for the coefficients.

Method

Consider the second-order linear differential equation:

$$a_2(z)f''(z) + a_1(z)f'(z) + a_0(z)f(z) = 0.$$

Suppose a_2 is nonzero for all z. Then we can divide throughout to obtain:

$$f'' + \frac{a_1(z)}{a_2(z)}f' + \frac{a_0(z)}{a_2(z)}f = 0.$$

Suppose further that a_1/a_2 and a_0/a_2 are analytic functions.

The power series method calls for the construction of a power series solution:

$$f = \sum_{k=0}^{\infty} A_k z^k.$$

If a_2 is zero for some z, then the Frobenius method, a variation on this method, is suited to deal with so called singular points. The method works analogously for higher order equations as well as for systems.

Let us look at the Hermite differential equation:

$$f'' - 2zf' + \lambda f = 0; \lambda = 1$$

We can try to construct a series solution:

$$f = \sum_{k=0}^{\infty} A_k z^k$$

$$f' = \sum_{k=1}^{\infty} k A_k z^{k-1}$$

$$f'' = \sum_{k=2}^{\infty} k(k-1) A_k z^{k-2}$$

Substituting these in the differential equation:

$$\sum_{k=2}^{\infty} k(k-1) A_k z^{k-2} - 2z \sum_{k=1}^{\infty} k A_k z^{k-1} + \sum_{k=0}^{\infty} A_k z^k = 0$$

$$= \sum_{k=2}^{\infty} k(k-1)A_k z^{k-2} - \sum_{k=1}^{\infty} 2kA_k z^k + \sum_{k=0}^{\infty} A_k z^k$$

Making a shift on the first sum:

$$= \sum_{k=0}^{\infty} (k+2)(k+1)A_{k+2} z^k - \sum_{k=1}^{\infty} 2kA_k z^k + \sum_{k=0}^{\infty} A_k z^k$$

$$= 2A_2 + \sum_{k=1}^{\infty} (k+2)(k+1)A_{k+2} z^k - \sum_{k=1}^{\infty} 2kA_k z^k + A_0 + \sum_{k=1}^{\infty} A_k z^k$$

$$= 2A_2 + A_0 + \sum_{k=1}^{\infty} \left((k+2)(k+1)A_{k\ 2} + (-2k+1)A_k \right) z$$

If this series is a solution, then all these coefficients must be zero, so for both k=0 and k>0:

$$(k+2)(k+1)A_{k+2} + (-2k+1)A_k = 0$$

We can rearrange this to get a recurrence relation for A_{k+2}.

$$(k+2)(k+1)A_{k+2} = -(-2k+1)A_k$$

$$A_{k+2} = \frac{(2k-1)}{(k+2)(k+1)} A_k$$

Now, we have,

$$A_2 = \frac{-1}{(2)(1)} A_0 = \frac{-1}{2} A_0, A_3 = \frac{1}{(3)(2)} A_1 = \frac{1}{6} A_1$$

We can determine A_0 and A_1 if there are initial conditions, i.e. if we have an initial value problem.

So we have,

$$A_4 = \frac{1}{4} A_2 = \left(\frac{1}{4} \right)\left(\frac{-1}{2} \right) A_0 = \frac{-1}{8} A_0$$

$$A_5 = \frac{1}{4} A_3 = \left(\frac{1}{4} \right)\left(\frac{1}{6} \right) A_1 = \frac{1}{24} A_1$$

$$A_6 = \frac{7}{30} A_4 = \left(\frac{7}{30} \right)\left(\frac{-1}{8} \right) A_0 = \frac{-7}{240} A_0$$

$$A_7 = \frac{3}{14} A_5 = \left(\frac{3}{14} \right)\left(\frac{1}{24} \right) A_1 = \frac{1}{112} A_1$$

and the series solution is:

$$f = A_0 z^0 + A_1 z^1 + A_2 z^2 + A_3 z^3 + A_4 z^4 + A_5 z^5 + A_6 z^6 + A_7 z^7 + \cdots$$

$$= A_0 z^0 + A_1 z^1 + \frac{-1}{2} A_0 z^2 + \frac{1}{6} A_1 z^3 + \frac{-1}{8} A_0 z^4 + \frac{1}{24} A_1 z^5 + \frac{-7}{240} A_0 z^6 + \frac{1}{112} A_1 z^7 + \cdots$$

$$= A_0 z^0 + \frac{-1}{2} A_0 z^2 + \frac{-1}{8} A_0 z^4 + \frac{-7}{240} A_0 z^6 + A_1 z + \frac{1}{6} A_1 z^3 + \frac{1}{24} A_1 z^5 + \frac{1}{112} A_1 z^7 + \cdots$$

which we can break up into the sum of two linearly independent series solutions:

$$f = A_0 \left(1 + \frac{-1}{2} z^2 + \frac{-1}{8} z^4 + \frac{-7}{240} z^6 + \cdots \right) + A_1 \left(z + \frac{1}{6} z^3 + \frac{1}{24} z^5 + \frac{1}{112} z^7 + \cdots \right)$$

which can be further simplified by the use of hypergeometric series.

Simpler Way using Taylor Series

A much simpler way of solving this equation (and power series solution in general) using the Taylor series form of the expansion. Here we assume the answer is of the form:

$$f = \sum_{k=0}^{\infty} \frac{A_k z^k}{k!}$$

If we do this, the general rule for obtaining the recurrence relationship for the coefficients is:

$$y^{[n]} \rightarrow A_{k+n}$$

and,

$$x^m y^{[n]} \rightarrow (k)(k-1)\cdots(k-m+1) A_{k+n-m}$$

In this case we can solve the Hermite equation in fewer steps:

$$f'' - 2zf' + \lambda f = 0; \lambda = 1$$

becomes,

$$A_{k+2} - 2k A_k + \lambda A_k = 0$$

or,

$$A_{k+2} = (2k - \lambda) A_k$$

in the series,

$$f = \sum_{k=0}^{\infty} \frac{A_k z^k}{k!}$$

Nonlinear Equations

The power series method can be applied to certain nonlinear differential equations, though with less flexibility. A very large class of nonlinear equations can be solved analytically by using the Parker–Sochacki method. Since the Parker–Sochacki method involves an expansion of the original system of ordinary differential equations through auxiliary equations, it is not simply referred to as the power series method. The Parker–Sochacki method is done before the power series method to make the power series method possible on many nonlinear problems. An ODE problem can be expanded with the auxiliary variables which make the power series method trivial for an equivalent, larger system. Expanding the ODE problem with auxiliary variables produces the same coefficients (since the power series for a function is unique) at the cost of also calculating the coefficients of auxiliary equations. Many times, without using auxiliary variables, there is no known way to get the power series for the solution to a system, hence the power series method alone is difficult to apply to most nonlinear equations.

The power series method will give solutions only to initial value problems (opposed to boundary value problems), this is not an issue when dealing with linear equations since the solution may turn up multiple linearly independent solutions which may be combined (by superposition) to solve boundary value problems as well. A further restriction is that the series coefficients will be specified by a nonlinear recurrence (the nonlinearities are inherited from the differential equation).

In order for the solution method to work, as in linear equations, it is necessary to express every term in the nonlinear equation as a power series so that all of the terms may be combined into one power series.

As an example, consider the initial value problem:

$$FF'' + 2F'^2 + \eta F' = 0 \quad ; \quad F(1) = 0 , F'(1) = -\frac{1}{2}$$

which describes a solution to capillary-driven flow in a groove. Note the two nonlinearities: the first and second terms involve products. Note also that the initial values are given at $\eta = 1$, which hints that the power series must be set up as:

$$F(\eta) = \sum_{i=0}^{\infty} c_i (\eta - 1)^i$$

since in this way,

$$\frac{d^n F}{d\eta^n}\bigg|_{\eta=1} = n! \, c_n$$

which makes the initial values very easy to evaluate. It is necessary to rewrite the equation slightly in light of the definition of the power series,

$$FF'' + 2F'^2 + (\eta - 1)F' + F' = 0 \quad ; \quad F(1) = 0 , F'(1) = -\frac{1}{2},$$

so that the third term contains the same form $\eta - 1$ that shows in the power series.

The last consideration is what to do with the products; substituting the power series in would result in products of power series when it's necessary that each term be its own power series. This is where the Cauchy product:

$$\left(\sum_{i=0}^{\infty} a_i x^i \right) \left(\sum_{i=0}^{\infty} b_i x^i \right) = \sum_{i=0}^{\infty} x^i \sum_{j=0}^{i} a_{i-j} b_j$$

is useful; substituting the power series into the differential equation and applying this identity leads to an equation where every term is a power series. After much rearrangement, the recurrence:

$$\sum_{j=0}^{i} \left((j+1)(j+2)c_{i-j}c_{j+2} + 2(i-j+1)(j+1)c_{i-j+1}c_{j+1} \right) + ic_i + (i+1)c_{i+1} = 0$$

is obtained, specifying exact values of the series coefficients. From the initial values, $c_0 = 0$ and $c_1 = -1/2$, thereafter the above recurrence is used. For example, the next few coefficients:

$$c_2 = -\frac{1}{6} \quad ; \quad c_3 = -\frac{1}{108} \quad ; \quad c_4 = \frac{7}{3240} \quad ; \quad c_5 = -\frac{19}{48600} \ldots$$

A limitation of the power series solution shows itself in this example. A numeric solution of the problem shows that the function is smooth and always decreasing to the left of $\eta = 1$, and zero to the right. At $\eta = 1$, a slope discontinuity exists, a feature which the power series is incapable of rendering, for this reason the series solution continues decreasing to the right of $\eta = 1$ instead of suddenly becoming zero.

Frobenius Method

In mathematics, the method of Frobenius, named after Ferdinand Georg Frobenius, is a way to find an infinite series solution for a second-order ordinary differential equation of the form:

$$z^2 u'' + p(z)zu' + q(z)u = 0$$

with,

$$u' \equiv \frac{du}{dz} \text{ and } u'' \equiv \frac{d^2 u}{dz^2}$$

in the vicinity of the regular singular point $z = 0$. One can divide by z^2 to obtain a differential equation of the form:

$$u'' + \frac{p(z)}{z} u' + \frac{q(z)}{z^2} u = 0$$

which will not be solvable with regular power series methods if either p(z)/z or q(z)/z² are not analytic at z = 0. The Frobenius method enables one to create a power series solution to such a differential equation, provided that p(z) and q(z) are themselves analytic at 0 or, being analytic elsewhere, both their limits at 0 exist (and are finite).

The method of Frobenius is to seek a power series solution of the form,

$$u(z) = \sum_{k=0}^{\infty} A_k z^{k+r}, \qquad (A_0 \neq 0)$$

Differentiating:

$$u'(z) = \sum_{k=0}^{\infty} (k+r) A_k z^{k+r-1}$$

$$u''(z) = \sum_{k=0}^{\infty} (k+r-1)(k+r) A_k z^{k+r-2}$$

Substituting:

$$z^2 \sum_{k=0}^{\infty} (k+r-1)(k+r) A_k z^{k+r-2} + z p(z) \sum_{k=0}^{\infty} (k+r) A_k z^{k+r-1} + q(z) \sum_{k=0}^{\infty} A_k z^{k+r}$$

$$= \sum_{k=0}^{\infty} (k+r-1)(k+r) A_k z^{k+r} + p(z) \sum_{k=0}^{\infty} (k+r) A_k z^{k+r} + q(z) \sum_{k=0}^{\infty} A_k z^{k+r}$$

$$= \sum_{k=0}^{\infty} \left[(k+r-1)(k+r) A_k z^{k+r} + p(z)(k+r) A_k z^{k+r} + q(z) A_k z^{k+r} \right]$$

$$= \sum_{k=0}^{\infty} \left[(k+r-1)(k+r) + p(z)(k+r) + q(z) \right] A_k z^{k+r}$$

$$= \left[r(r-1) + p(z)r + q(z) \right] A_0 z^r + \sum_{k=1}^{\infty} \left[(k+r-1)(k+r) + p(z)(k+r) + q(z) \right] A_k z^{k+r}$$

The expression,

$$r(r-1) + p(0)r + q(0) - I(r)$$

is known as the indicial polynomial, which is quadratic in r. The general definition of the indicial polynomial is the coefficient of the lowest power of z in the infinite series. In this case it happens to be that this is the rth coefficient but, it is possible for the lowest possible exponent to be r − 2, r − 1 or, something else depending on the given differential equation. This detail is important to keep in mind. In the process of synchronizing all the series of the differential equation to start at the same index value (which in the above expression is k = 1), one can end up with complicated expressions. However, in solving for the indicial roots attention is focused only on the coefficient of the lowest power of z.

Using this, the general expression of the coefficient of z^{k+r} is:

$$I(k+r) A_k + \sum_{j=0}^{k-1} \frac{(j+r) p^{(k-j)}(0) + q^{(k-j)}(0)}{(k-j)!} A_j,$$

These coefficients must be zero, since they should be solutions of the differential equation, so,

$$I(k+r)A_k + \sum_{j=0}^{k-1} \frac{(j+r)p^{(k-j)}(0)+q^{(k-j)}(0)}{(k-j)!}A_j = 0$$

$$\sum_{j=0}^{k-1} \frac{(j+r)p^{(k-j)}(0)+q^{(k-j)}(0)}{(k-j)!}A_j = -I(k+r)A_k$$

$$\frac{1}{-I(k+r)}\sum_{j=0}^{k-1} \frac{(j+r)p^{(k-j)}(0)+q^{(k-j)}(0)}{(k-j)!}A_j = A_k$$

The series solution with A_k above,

$$U_r(z) = \sum_{k=0}^{\infty} A_k z^{k+r}$$

satisfies,

$$z^2 U_r(z)'' + p(z)zU_r(z)' + q(z)U_r(z) = I(r)z^r$$

If we choose one of the roots to the indicial polynomial for r in $U_r(z)$, we gain a solution to the differential equation. If the difference between the roots is not an integer, we get another, linearly independent solution in the other root.

Let us solve,

$$z^2 f'' - zf' + (1-z)f = 0$$

Divide throughout by z² to give,

$$f'' - \frac{1}{z}f' + \frac{1-z}{z^2}f = f'' - \frac{1}{z}f' + \left(\frac{1}{z^2} - \frac{1}{z}\right)f = 0$$

which has the requisite singularity at z = 0.

Use the series solution,

$$f = \sum_{k=0}^{\infty} A_k z^{k+r} \quad f' = \sum_{k=0}^{\infty} (k+r)A_k z^{k+r-1} \quad f'' = \sum_{k=0}^{\infty} (k+r)(k+r-1)A_k z^{k+r-2}$$

Now, substituting,

$$\sum_{k=0}^{\infty}(k+r)(k+r)(k+r-1)A_k z^{k+r-2} - \frac{1}{z}\sum_{k=0}^{\infty}(k+r)A_k z^{k+r-1} + \left(\frac{1}{z^2} - \frac{1}{z}\right)\sum_{k=0}^{\infty} A_k z^{k+r}$$

$$= \sum_{k=0}^{\infty}(k+r)(k+r-1)A_k z^{k+r-2} - \frac{1}{z}\sum_{k=0}^{\infty}(k+r)A_k z^{k+r-1} + \frac{1}{z^2}\sum_{k=0}^{\infty} A_k z^{k+r} - \frac{1}{z}\sum_{k=0}^{\infty} A_k z^{k+r}$$

$$= \sum_{k=0}^{\infty}(k+r)(k+r-1)A_k z^{k+r-2} - \sum_{k=0}^{\infty}(k+r)A_k z^{k+r-2} + \sum_{k=0}^{\infty}A_k z^{k+r-2} - \sum_{k=0}^{\infty}A_k z^{k+r-1}$$

$$= \sum_{k=0}^{\infty}(k+r)(k+r-1)A_k z^{k+r-2} - \sum_{k=0}^{\infty}(k+r)A_k z^{k+r-2} + \sum_{k=0}^{\infty}A_k z^{k+r-2} - \sum_{k-1=0}^{\infty}A_{k-1} z^{k-1+r-1}$$

$$= \sum_{k=0}^{\infty}(k+r)(k+r-1)A_k z^{k+r-2} - \sum_{k=0}^{\infty}(k+r)A_k z^{k+r-2} + \sum_{k=0}^{\infty}A_k z^{k+r-2} - \sum_{k=1}^{\infty}A_{k-1} z^{k+r-2}$$

$$= \left\{ \sum_{k=0}^{\infty}\left((k+r)(k+r-1)-(k+r)+1\right)A_k z^{k+r-2}\right\} - \sum_{k=1}^{\infty}A_{k-1} z^{k+r-2}$$

$$= \left\{ \left(\;(\;-1)-\;+1\right)_0 {}^{r-2}+\sum_{k=1}^{\infty}\left((\;+\;)(\;+\;-1)-(\;+\;)+1\right)_k {}^{k+r-2}\right\} - \sum_{k=1}^{\infty}{}_{k-1}{}^{k+r-2}$$

$$= (r-1)^2 A_0 z^{r-2} + \left\{ \sum_{k=1}^{\infty}(k+r-1)^2 A_k z^{k+r-2} - \sum_{k=1}^{\infty}A_{k-1} z^{k+r-2}\right\}$$

$$= (r-1)^2 A_0 z^{r-2} + \sum_{k=1}^{\infty}\left((k+r-1)^2 A_k - A_{k-1}\right)z^{k+r-2}$$

From $(r-1)^2 = 0$ we get a double root of 1. Using this root, we set the coefficient of z^{k+r-2} to be zero (for it to be a solution), which gives us:

$$(k+1-1)^2 A_k - A_{k-1} = k^2 A_k - A_{k-1} = 0$$

hence we have the recurrence relation:

$$A_k = \frac{A_{k-1}}{k^2}$$

Given some initial conditions, we can either solve the recurrence entirely or obtain a solution in power series form.

Since the ratio of coefficients A_k / A_{k-1} is a rational function, the power series can be written as a generalized hypergeometric series.

Roots Separated by an Integer

The previous example involved an indicial polynomial with a repeated root, which gives only one solution to the given differential equation. In general, the Frobenius method gives two independent solutions provided that the indicial equation's roots are not separated by an integer (including zero).

If the root is repeated or the roots differ by an integer, then the second solution can be found using:

$$y_2 = C y_1 \ln x + \sum_{k=0}^{\infty}B_k x^{k+r_2}$$

where $y_1(x)$ is the first solution (based on the larger root in the case of unequal roots), r_2 is the smaller root, and the constant C and the coefficients B_k are to be determined. Once B_0 is chosen (for example by setting it to 1) then C and the B_k are determined up to but not including $B_{r_1-r_2}$, which can be set arbitrarily. This then determines the rest of the B_k In some cases the constant C must be zero. For example, consider the following differential equation (Kummer's equation with a = 1 and b = 2):

$$zu'' + (2-z)u' - u = 0$$

The roots of the indicial equation are –1 and 0. Two independent solutions are $1/z$ and $(e^z)/z$ so we see that the logarithm does not appear in any solution. The solution $(e^z-1)/z$ has a power series starting with the power zero. In a power series starting with z^{-1} the recurrence relation places no restriction on the coefficient for the term z^0 which can be set arbitrarily. If it is set to zero then with this differential equation all the other coefficients will be zero and we obtain the solution 1/z.

PARTIAL DIFFERENTIAL EQUATIONS

A partial differential equation (or briefly a PDE) is a mathematical equation that involves two or more independent variables, an unknown function (dependent on those variables), and partial derivatives of the unknown function with respect to the independent variables. The order of a partial differential equation is the order of the highest derivative involved. A solution (or a particular solution) to a partial differential equation is a function that solves the equation or, in other words, turns it into an identity when substituted into the equation. A solution is called general if it contains all particular solutions of the equation concerned.

The term exact solution is often used for second- and higher-order nonlinear PDEs to denote a particular solution.

Partial differential equations are used to mathematically formulate, and thus aid the solution of, physical and other problems involving functions of several variables, such as the propagation of heat or sound, fluid flow, elasticity, electrostatics, electrodynamics, etc.

First-Order Partial Differential Equations

General form of first-order partial differential equation:

A first-order partial differential equation with nindependent variables has the general form:

$$F(x_1, x_2, \ldots, x_n, w, \frac{\partial w}{\partial x_1}, \frac{\partial w}{\partial x_2}, \ldots, \frac{\partial w}{\partial x_n}) = 0,$$

where $w = w(x_1, x_2, \ldots, x_n)$ is the unknown function and F(...) is a given function.

Quasilinear Equations

General form of first-order quasilinear PDE:

A first-order quasilinear partial differential equation with two independent variables has the general form:

$$f(x, y, w)\frac{\partial w}{\partial x} + g(x, y, w)\frac{\partial w}{\partial y} = h(x, y, w).$$

Such equations are encountered in various applications (continuum mechanics, gas dynamics, hydrodynamics, heat and mass transfer, wave theory, acoustics, multiphase flows, chemical engineering, etc).

If the functions f, g, and h are independent of the unknown w, then equation:

$$f(x, y, w)\frac{\partial w}{\partial x} + g(x, y, w)\frac{\partial w}{\partial y} = h(x, y, w) \text{ is called linear.}$$

The system of ordinary differential equations:

$$\frac{dx}{f(x, y, w)} = \frac{dy}{g(x, y, w)} = \frac{dw}{h(x, y, w)}$$

is known as the characteristic system of equation $f(x, y, w)\frac{\partial w}{\partial x} + g(x, y, w)\frac{\partial w}{\partial y} = h(x, y, w)$. Suppose that two independent particular solutions of this system have been found in the form:

$$u_1(x, y, w) = C_1, \qquad u_2(x, y, w) = C_2,$$

where C_1 and C_2 are arbitrary constants; such particular solutions are known as integrals of system.

Then the general solution to equation $f(x, y, w)\frac{\partial w}{\partial x} + g(x, y, w)\frac{\partial w}{\partial y} = h(x, y, w)$. can be written as:

$$\Phi(u_1, u_2) = 0,$$

where Φ is an arbitrary function of two variables. With equation $\Phi(u_1, u_2) = 0$, solved for u2, one often specifies the general solution in the form $u_2 = \Psi(u_1)$, *where* $\Psi(u)$ is an arbitrary function of one variable.

If h(x,y,w)≡0, then w=C_2 can be used as the second integral in example. Consider the linear equation:

$$\frac{\partial w}{\partial x} + a\frac{\partial w}{\partial y} = b.$$

The associated characteristic system of ordinary differential equations,

$$\frac{dx}{1} = \frac{dy}{a} = \frac{dw}{b}$$

has two integrals,

$$y - ax = C_1, \qquad w - bx = C_2.$$

Therefore, the general solution to this PDE can be written as $w - bx = \Psi(y - ax)$, or

$$w = bx + \Psi(y - ax),$$

where $\Psi(z)$ is an arbitrary function.

Cauchy Problem

Generalized Cauchy problem: find a solution $w = w(x, y)$ to equation:

$$f(x, y, w)\frac{\partial w}{\partial x} + g(x, y, w)\frac{\partial w}{\partial y} = h(x, y, w) \text{ satisfying the initial conditions}$$

$$x = \varphi_1(\xi), \quad y = \varphi_2(\xi), \quad w = \varphi_3(\xi),$$

where ξ is a parameter ($\alpha \le \xi \le \beta$) and the $\varphi_k(\xi)$ are given functions.

Geometric interpretation: find an integral surface of equation:

$$f(x, y, w)\frac{\partial w}{\partial x} + g(x, y, w)\frac{\partial w}{\partial y} = h(x, y, w)$$

passing through the line defined parametrically by equation $x = \varphi_1(\xi), \quad y = \varphi_2(\xi), \quad w = \varphi_3(\xi)$.

Classical Cauchy problem: find a solution $w = w(x, y)$ of equation $f(x, y, w)\frac{\partial w}{\partial x} + g(x, y, w)\frac{\partial w}{\partial y} = h(x, y, w)$. satisfying the initial condition:

$$w = \varphi(y) \text{ at } x = 0$$

where $\varphi(y)$ is a given function.

It is often convenient to represent the classical Cauchy problem as a generalized Cauchy problem by rewriting condition $w = \varphi(y)$ at $x = 0$ in the parametric form:

$$x = 0, \ y = \xi, \ w = \varphi(\xi).$$

Existence and uniqueness theorem:

If the coefficients f, g, and h of equation $f(x, y, w)\frac{\partial w}{\partial x} + g(x, y, w)\frac{\partial w}{\partial y} = h(x, y, w)$. and the functions φ_k in are continuously differentiable with respect to each of their arguments and if the inequalities $f\varphi_2' - g\varphi_1' \ne 0$ and $(\varphi_1')^2 + (\varphi_2')^2 \ne 0|$ hold along the curve, then there is a unique solution to the Cauchy problem.

Procedure of Solving the Cauchy Problem

The procedure for solving the Cauchy problem involves several steps. First, two independent integrals of the characteristic system are determined. Then, to find the constants of integration C_1 and C_2, the initial data must be substituted into the integrals to obtain:

$$u_1(\varphi_1(\xi),\varphi_2(\xi),\varphi_3(\xi)) = C_1, \qquad u_2(\varphi_1(\xi),\varphi_2(\xi),\varphi_3(\xi)) = C_2.$$

Eliminating C_1 and C_2 from and yields:

$$u_1(x,y,w) = u_1(\varphi_1(\xi),\varphi_2(\xi),\varphi_3(\xi)),$$
$$u_2(x,y,w) = u_2(\varphi_1(\xi),\varphi_2(\xi),\varphi_3(\xi)).$$

Formulas are a parametric form of the solution to the Cauchy problem. In some cases, one may succeed in eliminating the parameter ξ from relations, thus obtaining the solution in an explicit form.

In the cases where first integrals of the characteristic system cannot be found using analytical methods, one should employ numerical methods to solve the Cauchy problem.

Second-order Partial Differential Equations

Linear, Semilinear and Nonlinear Second-order PDEs

A second-order linear partial differential equation with two independent variables has the form:

$$a(x,y)\frac{\partial^2 w}{\partial x^2} + 2b(x,y)\frac{\partial^2 w}{\partial x \partial y} + c(x,y)\frac{\partial^2 w}{\partial y_2} = \alpha(x,y)\frac{\partial w}{\partial x} + \beta(x,y)\frac{\partial w}{\partial y} + \gamma(x,y)w + \delta(x,y).$$

If $\delta \equiv 0$, equation:

$$a(x,y)\frac{\partial^2 w}{\partial x^2} + 2b(x,y)\frac{\partial^2 w}{\partial x \partial y} + c(x,y)\frac{\partial^2 w}{\partial y_2} = \alpha(x,y)\frac{\partial w}{\partial x} + \beta(x,y)\frac{\partial w}{\partial y} + \gamma(x,y)w + \delta(x,y).$$

is a homogeneous linear equation, and if $\delta \neq 0$, it is a nonhomogeneous linear equation. The functions a(x,y), b(x,y),... , γ(x,y), δ(x,y) are called coefficients of equation:

$$a(x,y)\frac{\partial^2 w}{\partial x^2} + 2b(x,y)\frac{\partial^2 w}{\partial x \partial y} + c(x,y)\frac{\partial^2 w}{\partial y_2} = \alpha(x,y)\frac{\partial w}{\partial x} + \beta(x,y)\frac{\partial w}{\partial y} + \gamma(x,y)w + \delta(x,y).$$

Some properties of a homogeneous linear equation (with $\delta \equiv 0$):

1. A homogeneous linear equation has a particular solution w = 0.

2. The principle of linear superposition holds; namely, if $w_1(x,y), w_2(x,y), ..., w_n(x,y)$ are particular solutions to homogeneous linear equation, then the function $A_1 w_1(x,y) + A_2 w_2(x,y) + \cdots + A_n w_n(x,y)$ where $A_1, A_2, ..., A$ are arbitrary numbers is also an exact solution to that equation.

3. Suppose equation:

$$a(x,y)\frac{\partial^2 w}{\partial x^2}+2b(x,y)\frac{\partial^2 w}{\partial x\partial y}+c(x,y)\frac{\partial^2 w}{\partial y_2}=\alpha(x,y)\frac{\partial w}{\partial x}+\beta(x,y)\frac{\partial w}{\partial y}+\gamma(x,y)w+\delta(x,y).$$

has a particular solution $\tilde{w}=\tilde{w}(x,y;\mu)$ that depends on a parameter μ, and the coefficients of the linear differential equation are independent of μ (but can depend on x and y).

Then, by differentiating $\tilde{w}$ with respect to μ, one obtains other solutions to the equation,

$$\frac{\partial \overline{w}}{\partial \mu},\quad \frac{\partial^2 \overline{w}}{\partial \mu^2},\quad \ldots,\quad \frac{\partial^k \overline{w}}{\partial \mu^k},\quad \ldots$$

4. Let $\tilde{w}=\tilde{w}(x,y;\mu)$ be a particular solution as described in property 3. Multiplying $\tilde{w}$ by an arbitrary function $\varphi(\mu)$ and integrating the resulting expression with respect to μ over some interval $[\mu_1,\mu_2]$, one obtains a new function $\int_{\mu_1}^{\mu_2}\tilde{w}(x,y;\mu)\varphi(\mu)d\mu$, which is also a solution to the original homogeneous linear equation.

5. Suppose the coefficients of the homogeneous linear equation

$$a(x,y)\frac{\partial^2 w}{\partial x^2}+2b(x,y)\frac{\partial^2 w}{\partial x\partial y}+c(x,y)\frac{\partial^2 w}{\partial y_2}=\alpha(x,y)\frac{\partial w}{\partial x}+\beta(x,y)\frac{\partial w}{\partial y}+\gamma(x,y)w+\delta(x,y).$$

are independent of x. Then: (i) there is a particular solution of the form w=e^{\lambda x-}u(y), where λ is an arbitrary number and u(y) is determined by a linear second-order ordinary differential equation, and (ii) differentiating any particular solution with respect to x also results in a particular solution to equation

$$a(x,y)\frac{\partial^2 w}{\partial x^2}+2b(x,y)\frac{\partial^2 w}{\partial x\partial y}+c(x,y)\frac{\partial^2 w}{\partial y_2}=\alpha(x,y)\frac{\partial w}{\partial x}+\beta(x,y)\frac{\partial w}{\partial y}+\gamma(x,y)w+\delta(x,y).$$

Properties 2–5 are widely used for constructing solutions to problems governed by linear PDEs.

Semilinear and Nonlinear Second-order PDEs

A second-order semilinear partial differential equation with two independent variables has the form:

$$a(x,y)\frac{\partial^2 w}{\partial x^2}+2b(x,y)\frac{\partial^2 w}{\partial x\partial y}+c(x,y)\frac{\partial^2 w}{\partial y^2}=F\left(x,y,w,\frac{\partial w}{\partial x},\frac{\partial w}{\partial x}\right).$$

In the general case, a second-order nonlinear partial differential equation with two independent variables has the form:

$$F(x,y,w,\frac{\partial w}{\partial x},\frac{\partial w}{\partial y},\frac{\partial^2 w}{\partial x^2},\frac{\partial^2 w}{\partial x\partial y},\frac{\partial^2 w}{\partial y^2})=0.$$

Some Linear Equations Encountered in Applications

Three basic types of linear partial differential equations are distinguished—parabolic, hyperbolic, and elliptic. The solutions of the equations pertaining to each of the types have their own characteristic qualitative differences.

Heat Equation (A Parabolic Equation)

1. The simplest example of a parabolic equation is the heat equation:

$$\frac{\partial w}{\partial t} - \frac{\partial^2 w}{\partial x^2} = 0,$$

where the variables t and x play the role of time and a spatial coordinate, respectively. Note that equation $\frac{\partial w}{\partial t} - \frac{\partial^2 w}{\partial x^2} = 0$, contains only one highest derivative term.

Equation $\frac{\partial w}{\partial t} - \frac{\partial^2 w}{\partial x^2} = 0$, is often encountered in the theory of heat and mass transfer. It describes one-dimensional unsteady thermal processes in quiescent media or solids with constant thermal diffusivity. A similar equation is used in studying corresponding one-dimensional unsteady mass-exchange processes with constant diffusivity.

2. The heat equation $\frac{\partial w}{\partial t} - \frac{\partial^2 w}{\partial x^2} = 0$, has infinitely many particular solutions (this fact is common to many PDEs); in particular, it admits solutions:

$$w(x,t) = A(x^2 + 2t) + B,$$
$$w(x,t) = A\exp(\mu^2 t \pm \mu x) + B,$$
$$w(x,t) = A\frac{1}{\sqrt{t}}\exp\left(-\frac{x^2}{4t}\right) + B,$$
$$w(x,t) = A\exp(-\mu^2 t)\cos(\mu x + B) + C,$$
$$w(x,t) = A\exp(-\mu x)\cos(\mu x - 2\mu^2 t + B) + C,$$

where A, B, C, and μ are arbitrary constants.

Wave Equation (A Hyperbolic Equation)

1. The simplest example of a hyperbolic equation is the wave equation:

$$\frac{\partial^2 w}{\partial t^2} - \frac{\partial^2 w}{\partial x^2} = 0,$$

where the variables t and x play the role of time and the spatial coordinate, respectively. Note that the highest derivative terms in equation $\frac{\partial^2 w}{\partial t^2} - \frac{\partial^2 w}{\partial x^2} = 0$, differ in sign.

This equation is also known as the equation of vibration of a string. It is often encountered in elasticity, aerodynamics, acoustics, and electrodynamics.

2. The general solution of the wave equation $\dfrac{\partial^2 w}{\partial t^2} - \dfrac{\partial^2 w}{\partial x^2} = 0$, is:

$$w = \varphi(x+t) + \psi(x-t),$$

Where $\varphi(x)$ and $\psi(x)$ are arbitrary twice continuously differentiable functions. This solution has the physical interpretation of two traveling waves of arbitrary shape that propagate to the right and to the left along the x-axis with a constant speed equal to 1.

Laplace Equation (An Elliptic Equation)

1. The simplest example of an elliptic equation is the Laplace equation:

$$\frac{\partial^2 w}{\partial x^2} + \frac{\partial^2 w}{\partial y^2} = 0,$$

where x and y play the role of the spatial coordinates. Note that the highest derivative terms in equation $\dfrac{\partial^2 w}{\partial x^2} + \dfrac{\partial^2 w}{\partial y^2} = 0$, have like signs. The Laplace equation is often written briefly as $\Delta w = 0$, where Δ is the Laplace operator.

The Laplace equation is often encountered in heat and mass transfer theory, fluid mechanics, elasticity, electrostatics, and other areas of mechanics and physics. For example, in heat and mass transfer theory, this equation describes steady-state temperature distribution in the absence of heat sources and sinks in the domain under study.

A solution to the Laplace equation $\dfrac{\partial^2 w}{\partial x^2} + \dfrac{\partial^2 w}{\partial y^2} = 0$, is called a harmonic function.

2. Note some particular solutions of the Laplace equation $\dfrac{\partial^2 w}{\partial x^2} + \dfrac{\partial^2 w}{\partial y^2} = 0,$:

$$w(x, y) = Ax + By + C,$$

$$w(x, y) = A(x^2 - y^2) + Bxy,$$

$$w(x, y) = \frac{Ax + By}{x^2 + y^2} + C,$$

$$w(x, y) = (A \sinh \mu x + B \cosh \mu x)(C \cos \mu y + D \sin \mu y),$$

$$w(x, y) = (A \cos \mu x + B \sin \mu x)(C \sinh \mu y + D \cosh \mu y),$$

where A, B, C, D, and μ are arbitrary constants.

A fairly general method for constructing solutions to the Laplace equation $\dfrac{\partial^2 w}{\partial x^2} + \dfrac{\partial^2 w}{\partial y^2} = 0$, involves the following. Let $f(z) = u(x, y) + iv(x, y)$ be any analytic function of the complex variable $z = x + iy$ (u and v are real functions of the real variables x and y ; $i^2 = -1$). Then the real and imaginary parts of f both satisfy the Laplace equation,

$$\Delta u = 0, \quad \Delta v = 0.$$

Thus, by specifying analytic functions f (z) and taking their real and imaginary parts, one obtains various solutions of the Laplace equation $\dfrac{\partial^2 w}{\partial x^2}+\dfrac{\partial^2 w}{\partial y^2}=0.$.

Classification of Second-order Partial Differential Equations

Types of Equations

Any semilinear partial differential equation of the second-order with two independent variables
$$a(x,y)\frac{\partial^2 w}{\partial x^2}+2b(x,y)\frac{\partial^2 w}{\partial x\partial y}+c(x,y)\frac{\partial^2 w}{\partial y^2}=F\left(x,y,w,\frac{\partial w}{\partial x},\frac{\partial w}{\partial x}\right)\cdot$$ can be reduced, by appropriate
manipulations, to a simpler equation that has one of the three highest derivative combinations specified above in examples $\dfrac{\partial w}{\partial t}-\dfrac{\partial^2 w}{\partial x^2}=0,\ \dfrac{\partial^2 w}{\partial t^2}-\dfrac{\partial^2 w}{\partial x^2}=0,$ and $\dfrac{\partial^2 w}{\partial x^2}+\dfrac{\partial^2 w}{\partial y^2}=0.$

Given a point $(x,y)|$, equation $a(x,y)\dfrac{\partial^2 w}{\partial x^2}+2b(x,y)\dfrac{\partial^2 w}{\partial x\partial y}+c(x,y)\dfrac{\partial^2 w}{\partial y^2}=F\left(x,y,w,\dfrac{\partial w}{\partial x},\dfrac{\partial w}{\partial x}\right).$ is said to be:

> parabolic   if $b^2-ac=0,$
>
> hyperbolic if $b^2-ac>0,$
>
> elliptic   if $b^2-ac<0$

at this point.

Characteristic Equations

In order to reduce equation $a(x,y)\dfrac{\partial^2 w}{\partial x^2}+2b(x,y)\dfrac{\partial^2 w}{\partial x\partial y}+c(x,y)\dfrac{\partial^2 w}{\partial y^2}=F\left(x,y,w,\dfrac{\partial w}{\partial x},\dfrac{\partial w}{\partial x}\right).$ to a canonical form, one should first write out the characteristic equation:

$$a(dy)^2-2b\,dx\,dy+c(dx)^2=0,$$

which with a $\neq 0$ splits into two equations:

$$a\,dy-(b+\sqrt{b^2-ac})\,dx=0$$

and

$$a\,dy-(b-\sqrt{b^2-ac})\,dx=0$$

and then find their general integrals.

If a≡o, the simpler equations:

$$dx=0,$$
$$2b\,dy-c\,dx=0$$

should be used instead of $a\,dy-(b+\sqrt{b^2-ac})\,dx=0$ and $a\,dy-(b-\sqrt{b^2-ac})\,dx=0$. The first equation has the obvious general solution x=C.

Canonical Form of Parabolic Equations (Case b²–ac=0)

In this case, equations $a\,dy-(b+\sqrt{b^2-ac})\,dx=0$ and $a\,dy-(b-\sqrt{b^2-ac})\,dx=0$ coincide and have a common general integral,

$$u(x,y)=C.$$

By passing from x, y to new independent variables ξ, η in accordance with the relations:

$$\xi=u(x,y), \qquad \eta=\eta(x,y),$$

where $\eta=\eta(x,y)$ is any twice differentiable function that satisfies the condition of nondegeneracy of the Jacobian $\dfrac{D(\xi,\eta)}{D(x,y)}$ in the given domain, one reduces equation

$$a(x,y)\frac{\partial^2 w}{\partial x^2}+2b(x,y)\frac{\partial^2 w}{\partial x\partial y}+c(x,y)\frac{\partial^2 w}{\partial y^2}=F\left(x,y,w,\frac{\partial w}{\partial x},\frac{\partial w}{\partial x}\right).$$ to the canonical form:

$$\frac{\partial^2 w}{\partial \eta^2}=F_1\left(\xi,\eta,w,\frac{\partial w}{\partial \xi},\frac{\partial w}{\partial \eta}\right).$$

As η, one can take η=x or η=y .

It is apparent that the transformed equation $\dfrac{\partial^2 w}{\partial \eta^2}=F_1\left(\xi,\eta,w,\dfrac{\partial w}{\partial \xi},\dfrac{\partial w}{\partial \eta}\right).$ has only one highest-de-

rivative term, just as the heat equation $\dfrac{\partial w}{\partial t}-\dfrac{\partial^2 w}{\partial x^2}=0,$

Two canonical forms of hyperbolic equations (case b² – ac > 0):

1. The general integrals:

$$u_1(x,y)=C_1, \qquad u_2(x,y)=C_2$$

of equations $a\,dy-(b+\sqrt{b^2-ac})\,dx=0$ and $a\,dy-(b-\sqrt{b^2-ac})\,dx=0$ are real and different. These integrals determine two different families of real characteristics.

By passing from x, y to new independent variables ξ, η in accordance with the relations:

$$\xi=u_1(x,y), \qquad \eta=u_2(x,y),$$

one reduces equation $a(x,y)\dfrac{\partial^2 w}{\partial x^2}+2b(x,y)\dfrac{\partial^2 w}{\partial x\partial y}+c(x,y)\dfrac{\partial^2 w}{\partial y^2}=F\left(x,y,w,\dfrac{\partial w}{\partial x},\dfrac{\partial w}{\partial x}\right)$ to:

$$\frac{\partial^2 w}{\partial \xi\partial \eta}=F_2\left(\xi,\eta,w,\frac{\partial w}{\partial \xi},\frac{\partial w}{\partial \eta}\right).$$

This is the so-called first canonical form of a hyperbolic equation.

2. The transformation:

$$\xi = t + z, \quad \eta = t - z$$

brings the above equation to another canonical form,

$$\frac{\partial^2 w}{\partial t^2} - \frac{\partial^2 w}{\partial z^2} = F_3\left(t, z, w, \frac{\partial w}{\partial t}, \frac{\partial w}{\partial z}\right),$$

where $F_3 = 4\,F_2$. This is the so-called second canonical form of a hyperbolic equation. Apart from notation, the left-hand side of the last equation coincides with that of the wave equation $\frac{\partial^2 w}{\partial t^2} - \frac{\partial^2 w}{\partial x^2} = 0$.

Canonical Form of Elliptic Equations (Case b² – ac < 0)

In this case the general integrals of equations $a\,dy - (b + \sqrt{b^2 - ac})\,dx = 0$ and $a\,dy - (b - \sqrt{b^2 - ac})\,dx = 0$ are complex conjugates; these determine two families of complex characteristics.

Let the general integral of equation $a\,dy - (b + \sqrt{b^2 - ac})\,dx = 0$ have the form:

$$u_1(x, y) + iu_2(x, y) = C, \quad i^2 = -1,$$

where u_1(x,y) and u_2(x,y) are real-valued functions.

By passing from x, y to new independent variables ξ, η in accordance with the relations,

$$\xi = u_1(x, y) \qquad \eta = u_2 \; x \; y \;,$$

one reduces equation $a(x, y)\frac{\partial^2 w}{\partial x^2} + 2b(x, y)\frac{\partial^2 w}{\partial x \partial y} + c(x, y)\frac{\partial^2 w}{\partial y^2} = F\left(x, y, w, \frac{\partial w}{\partial x}, \frac{\partial w}{\partial x}\right)$.to the canonical form,

$$\frac{\partial^2 w}{\partial \xi^2} + \frac{\partial^2 w}{\partial \eta^2} = F_4\left(\xi, \eta, w, \frac{\partial w}{\partial \xi}, \frac{\partial w}{\partial \eta}\right).$$

Apart from notation, the left-hand side of the last equation coincides with that of the Laplace equation $\frac{\partial^2 w}{\partial x^2} + \frac{\partial^2 w}{\partial y^2} = 0$.

Basic Problems for PDEs of Mathematical Physics

Most PDEs of mathematical physics govern infinitely many qualitatively similar phenomena or processes. This follows from the fact that differential equations have, as a rule, infinitely many particular solutions. The specific solution that describes the physical phenomenon under study is separated from the set of particular solutions of the given differential equation by means of the initial and boundary conditions.

For simplicity and clarity of illustration, the basic problems of mathematical physics will be presented for the simplest linear equations $\dfrac{\partial w}{\partial t} - \dfrac{\partial^2 w}{\partial x^2} = 0$, $\dfrac{\partial^2 w}{\partial t^2} - \dfrac{\partial^2 w}{\partial x^2} = 0$, and $\dfrac{\partial^2 w}{\partial x^2} + \dfrac{\partial^2 w}{\partial y^2} = 0$, only

Cauchy problem and boundary value problems for parabolic equations.

Cauchy problem $(t \geq 0, -\infty < x < \infty)$. Find a function w that satisfies heat equation $\dfrac{\partial w}{\partial t} - \dfrac{\partial^2 w}{\partial x^2} = 0$, for t>o and the initial condition:

$$w = \varphi(x) \text{ at } t = 0.$$

The solution of the Cauchy problem is,

$$w(x,t) = \int_{-\infty}^{\infty} \varphi(\xi) E(x, \xi, t) d\xi,$$

where E(x,ξ,t) is the fundamental solution of the Cauchy problem,

$$E(x, \xi, t) = \frac{1}{2\sqrt{\pi a t}} \exp\left[-\frac{(x - \xi)^2}{4at} \right].$$

In all boundary value problems (or initial-boundary value problems) below, it will be required to find a function w, in a domain $t \geq 0$, $x1 \leq x \leq x2$ $(-\infty < x1 < x2 < \infty)$, that satisfies the heat equation $\dfrac{\partial w}{\partial t} - \dfrac{\partial^2 w}{\partial x^2} = 0$, for $t > 0$ and the initial condition. In addition, all problems will be supplemented with some boundary conditions as given below.

First boundary value problem. The function w(x,t) takes prescribed values on the boundary:

$$w = \psi_1(t) \quad \text{at} \quad x = x_1,$$
$$w = \psi_2(t) \quad \text{at} \quad x = x_2.$$

In particular, the solution to the first boundary value problem $\dfrac{\partial w}{\partial t} - \dfrac{\partial^2 w}{\partial x^2} = 0$, $w = \varphi(x)$ at $t = 0$. $w = \psi_2(t)$ at $x = x_2$ with $\psi_1(t) = \psi_2(t) \equiv 0$, $x_1 = 0$, and $x_2 = l$ is expressed as:

$$w(x,t) = \int_0^l \varphi(\xi) G(x, \xi, t) d\xi,$$

where the Green's function G(x,ξ,t) is defined by the formulas:

$$G(x, \xi, t) = \frac{2}{l} \sum_{n=1}^{\infty} \sin\left(\frac{n\pi x}{l}\right) \sin\left(\frac{n\pi \xi}{l}\right) \exp\left(-\frac{an^2\pi^2 t}{l^2}\right)$$

$$= \frac{1}{2\sqrt{\pi a t}} \sum_{n=-\infty}^{\infty} \left\{ \exp\left[-\frac{(x + \xi + 2nl)^2}{4at} \right] - \exp\left[-\frac{(x + \xi + 2nl)^2}{4at} \right] \right\}.$$

The first series converges rapidly at large t and the second series at small t.

Second boundary value problem. The derivatives of the function w(x,t) are prescribed on the boundary:

$$\frac{\partial w}{\partial x} = \psi_1(t) \quad \text{at} \quad x = x_1,$$

$$\frac{\partial w}{\partial x} = \psi_2(t) \quad \text{at} \quad x = x_2,$$

Third boundary value problem. A linear relationship between the unknown function and its derivatives are prescribed on the boundary:

$$\frac{\partial w}{\partial x} - k_1 w = \psi_1(t) \quad \text{at} \quad x = x_1,$$

$$\frac{\partial w}{\partial x} - k_2 w = \psi_2(t) \quad \text{at} \quad x = x_2.$$

Mixed boundary value problems. Conditions of different type, listed above, are set on the boundary of the domain in question, for example,

$$x = \psi_1(t) \text{ at} \quad x = x_1,$$

$$\frac{\partial w}{\partial x} = \psi_2(t) \quad \text{at} \quad x = x_2.$$

The boundary conditions above four equations are called homogeneous if $\psi1(t)=\psi2(t)\equiv0$.

Solutions to the above initial-boundary value problems for the heat equation can be obtained by separation of variables(Fourier method) in the form of infinite series or by the method of integral transforms using the Laplace transform.

Cauchy Problem and Boundary Value Problems for Hyperbolic Equations

Cauchy problem $(t \geq 0, -\infty < x < \infty)$.Find a function w that satisfies the wave equation $\frac{\partial^2 w}{\partial t^2} - \frac{\partial^2 w}{\partial x^2} = 0$, for t>0 and two initial conditions:

$$w = \varphi_0(x) \text{ at } t = 0,$$

$$\frac{\partial w}{\partial t} = \varphi_1(x) \text{at } t = 0.$$

The solution of the Cauchy problem is given by D'Alembert's formula:

$$w(x,t) = \frac{1}{2}[\varphi_0(x+at) + \varphi_0(x-at)] + \frac{1}{2a}\int_{x-at}^{x+at} \varphi_1(\xi)d\xi.$$

Boundary value problems. In all boundary value problems, it is required to find a function w, in a domain $t \geq 0, x_1 \leq x \leq x_2 (-\infty < x_1 < x_2 < \infty)$, that satisfies the wave equation $\dfrac{\partial^2 w}{\partial t^2} - \dfrac{\partial^2 w}{\partial x^2} = 0$, for t>o and the initial conditions. In addition, appropriate boundary conditions, are imposed.

Solutions to these boundary value problems for the wave equation can be obtained by separation of variables (Fourier method) in the form of infinite series. In particular, the solution to the first boundary value problem with homogeneous boundary conditions, $\psi_1(t) = \psi_2(t) \equiv 0$ at $x_1 = 0$ and $x_2 = l$, is expressed as:

$$w(x,t) = \dfrac{\partial}{\partial t} \int_0^l \varphi_0(\xi) G(x,\xi,t)\, d\xi + \int_0^l \varphi_1(\xi) G(x,\xi,t)\, d\xi,$$

where,

$$G(x,\xi,t) = \dfrac{2}{a\pi} \sum_{n=1}^{\infty} \dfrac{1}{n} \sin\left(\dfrac{n\pi x}{l}\right) \sin\left(\dfrac{n\pi\xi}{l}\right) \sin\left(\dfrac{n\pi a t}{l}\right).$$

Goursat problem: On the characteristics of the wave equation $\dfrac{\partial^2 w}{\partial t^2} - \dfrac{\partial^2 w}{\partial x^2} = 0$, values of the unknown function w are prescribed:

$w = \varphi(x)$ for $x - t = 0$ $(0 \leq x \leq a)$,

$w = \psi(x)$ for $x + t = 0$ $(0 \leq x \leq b)$,

with the consistency condition $\varphi(0) = \psi(0)$ implied to hold.

Substituting the values set on the characteristics into the general solution of the wave equation $w = \varphi(x+t) + \psi(x-t),$, one arrives at a system of linear algebraic equations for $\varphi(x)$ and $\psi(x)$. As a result, the solution to the Goursat problem is obtained in the form:

$$w(x,t) = \varphi\left(\dfrac{x+t}{2}\right) + \psi\left(\dfrac{x-t}{2}\right) - \varphi(0).$$

The solution propagation domain is the parallelogram bounded by the four lines:

$$x - t = 0, x + t = 0, x - t = 2b, x + t = 2a.$$

Boundary Value Problems for Elliptic Equations

Setting boundary conditions for the first, second, and third boundary value problems for the Laplace equation $\dfrac{\partial^2 w}{\partial x^2} + \dfrac{\partial^2 w}{\partial y^2} = 0$, means prescribing values of the unknown function, its first derivative, and a linear combination of the unknown function and its derivative, respectively.

For example, the first boundary value problem in a rectangular domain o ≤ x ≤ a, o ≤ y ≤ b is characterized by the boundary conditions:

$w = \varphi_1(y)$ at $x = 0, w = \varphi_2(y)$ at $x = a,$

$w = \varphi_3(x)$ at $y = 0, w = \varphi_4(x)$ at $y = b.$

The solution to problem with $\varphi_3(x) = \varphi_4(x) \equiv 0$ is given by:

$$w(x, y) = \sum_{n=1}^{\infty} A_n \sinh\left[\frac{n\pi}{b}(a-x)\right]\sin\left(\frac{n\pi}{b}y\right) + \sum_{n=1}^{\infty} B_n \sinh\left(\frac{n\pi}{b}x\right)\sin\left(\frac{n\pi}{b}y\right),$$

where the coefficients A_n and B_n are expressed as:

$$A_n = \frac{2}{\lambda_n}\int_0^b \varphi_1(\xi)\sin\left(\frac{n\pi\xi}{b}\right)d\xi, \ \ B_n = \frac{2}{\lambda_n}\int_0^b \varphi_2(\xi)\sin\left(\frac{n\pi\xi}{b}\right)d\xi, \lambda_n = b\sinh\left(\frac{n\pi a}{b}\right).$$

For elliptic equations, the first boundary value problem is often called the Dirichlet problem, and the second boundary value problem is called the Neumann problem.

Some Nonlinear Equations Encountered in Applications

Nonlinear heat equation:

$$\frac{\partial w}{\partial t} = \frac{\partial}{\partial x}\left[f(w)\frac{\partial w}{\partial x}\right].$$

This equation describes one-dimensional unsteady thermal processes in quiescent media or solids in the case where the thermal diffusivity is temperature dependent, $f(w) > 0$. In the special case $f(w) \equiv 1$, the nonlinear equation $\frac{\partial w}{\partial t} = \frac{\partial}{\partial x}\left[f(w)\frac{\partial w}{\partial x}\right]$. becomes the linear heat equation $\frac{\partial w}{\partial t} - \frac{\partial^2 w}{\partial x^2} = 0$.

In general, the nonlinear heat equation $\frac{\partial w}{\partial t} = \frac{\partial}{\partial x}\left[f(w)\frac{\partial w}{\partial x}\right]$. admits exact solutions of the form:

$$w = W(kx - \lambda t) \quad \text{(traveling- wave solution)},$$
$$w = U(x/\sqrt{t}) \quad \text{(self- similar solution)},$$

Where $W = W(z)$ and $U = U(r)$ are determined by ordinary differential equations, and k and λ are arbitrary constants.

Kolmogorov–Petrovskii–Piskunov equation:

$$\frac{\partial w}{\partial t} = a\frac{\partial^2 w}{\partial x^2} + f(w), \quad a > 0.$$

Equations of this form are often encountered in various problems of mass and heat transfer (with f being the rate of a volume chemical reaction), combustion theory, biology, and ecology.

In the special case of f(w)≡0 and a=1, the nonlinear equation $\frac{\partial w}{\partial t} = a\frac{\partial^2 w}{\partial x^2} + f(w), \quad a > 0.$ becomes the linear heat equation $\frac{\partial w}{\partial t} - \frac{\partial^2 w}{\partial x^2} = 0,$

Equation $\dfrac{\partial w}{\partial t} = a\dfrac{\partial^2 w}{\partial x^2} + f(w), \quad a>0.$ is also called a heat equation with a nonlinear source.

Burgers equation:

$$\frac{\partial w}{\partial t} + w\frac{\partial w}{\partial x} = \frac{\partial^2 w}{\partial x^2}.$$

This equation is used for describing wave processes in gas dynamics, hydrodynamics, and acoustics.

1. Exact solutions to the Burgers equation can be obtained using the following formula (Hopf–Cole transformation):

$$w(x,t) = -\frac{2}{u}\frac{\partial u}{\partial x},$$

where $u=u(x,t)$ is a solution to the linear heat equation $u_t = u_{xx}$.

2. The solution to the Cauchy problem for the Burgers equation with the initial condition:

$$w = f(x) \quad \text{at} \quad t = 0 \;\; (-\infty < x < \infty)$$

has the form:

$$w(x,t) = -2\frac{\partial}{\partial x}\ln F(x,t),$$

where,

$$F(x,t) = \frac{1}{\sqrt{4\pi t}}\int_{-\infty}^{\infty}\exp\left[-\frac{(x-\xi)^2}{4t} + \frac{1}{2}\int_0^{\xi} f(\xi')d\xi'\right]d\xi.$$

Nonlinear wave equation:

$$\frac{\partial^2 w}{\partial t^2} = \frac{\partial}{\partial x}\left[f(w)\frac{\partial w}{\partial x}\right].$$

This equation is encountered in wave and gas dynamics, $f(w) > 0$. In the special case $f(w)\equiv 1$, the nonlinear equation $\dfrac{\partial^2 w}{\partial t^2} = \dfrac{\partial}{\partial x}\left[f(w)\dfrac{\partial w}{\partial x}\right]$. becomes the linear wave equation $\dfrac{\partial^2 w}{\partial t^2} - \dfrac{\partial^2 w}{\partial x^2} = 0,$

Equation $\dfrac{\partial^2 w}{\partial t^2} = \dfrac{\partial}{\partial x}\left[f(w)\dfrac{\partial w}{\partial x}\right]$. admits exact solutions in implicit form:

$$x + t\sqrt{f(w)} = \varphi(w),$$
$$x - t\sqrt{f(w)} = \psi(w),$$

Where $\varphi(w)$ and $\psi(w)$ are arbitrary functions.

Equation $\dfrac{\partial^2 w}{\partial t^2} = \dfrac{\partial}{\partial x}\left[f(w)\dfrac{\partial w}{\partial x}\right]$. can be reduced to a linear.

Nonlinear Klein–Gordon equation:

$$\frac{\partial^2 w}{\partial t^2} = a\frac{\partial^2 w}{\partial x^2} + f(w),\ a > 0.$$

Equations of this form arise in differential geometry and various areas of physics (superconductivity, dislocations in crystals, waves in ferromagnetic materials, laser pulses in two-phase media,

and others). For $f(w) \equiv 0$ and $a = 1$, equation $\dfrac{\partial^2 w}{\partial t^2} = a\dfrac{\partial^2 w}{\partial x^2} + f(w),\ a > 0$ coincides with the linear

wave equation $\dfrac{\partial^2 w}{\partial t^2} - \dfrac{\partial^2 w}{\partial x^2} = 0.$

1. In general, the nonlinear Klein–Gordon equation $\dfrac{\partial^2 w}{\partial t^2} = a\dfrac{\partial^2 w}{\partial x^2} + f(w),\ a > 0$ admits exact solutions of the form:

$$w = W(z),\quad z = kx - \lambda t,$$
$$w = U(\xi),\quad \xi = (\sqrt{at} + C_1)^2 - (x + C_2)^2,$$

where $W = W(z)$ and $U = U(\xi)$ are determined by ordinary differential equations, while $k,\ \lambda, C_1,$ and C_2 are arbitrary constants.

2. In the special case:

$$f(w) = be^{\beta w},$$

the general solution of equation $\dfrac{\partial^2 w}{\partial t^2} = a\dfrac{\partial^2 w}{\partial x^2} + f(w),\ a > 0$ is expressed as:

$$w(x,t) = \frac{1}{\beta}[\varphi(z) + \psi(y)] - \frac{2}{\beta}\ln\left|k\int \exp[\varphi(z)]dz - \frac{b\beta}{8ak}\int \exp[\psi(y)]dy\right|,$$

$$z = x - \sqrt{at},\qquad y = x + \sqrt{at},$$

where $\varphi = \varphi(z)$ and $\psi = \psi(y)$ are arbitrary functions and k is an arbitrary constant.

In the special cases $f(w) = b\sin(\beta w)$ and $f(w) = b\sinh(\beta w)$, equation is called the sine-Gordon equationand the sinh-Gordon equation, respectively.

Nonlinear Laplace equation:

$$\frac{\partial^2 w}{\partial x^2} + \frac{\partial^2 w}{\partial y^2} = f(w).$$

This equation is also called a stationary heat equation with a nonlinear source.

1. In general, the nonlinear heat equation $\dfrac{\partial^2 w}{\partial x^2} + \dfrac{\partial^2 w}{\partial y^2} = f(w)$. admits exact solutions of the form:

$$w = W(z), \quad z = k_1 x + k_2 y,$$
$$w = U(r), \quad r = \sqrt{(x + C_1)^2 + (y + C_2)^2},$$

where $W = W(z)$ and $U = U(r)|$ are determined by ordinary differential equations, while $k_1, k_2, C_1,$ and C_2 are arbitrary constants.

2. In the special case:

$$f(w) = ae^{\beta w},$$

the general solution of equation $\dfrac{\partial^2 w}{\partial x^2} + \dfrac{\partial^2 w}{\partial y^2} = f(w)$. is expressed as:

$$w(x, y) = -\frac{2}{\beta} \ln \frac{\left|1 - 2a\beta\Phi(z)\overline{\Phi(z)}\right|}{4\,|\,\Phi'_z(z)\,|},$$

where $\Phi = \Phi(z)$ is an arbitrary analytic function of the complex variable $z = x + iy$ with nonzero derivative, and the bar over a symbol denotes the complex conjugate.

Monge–Ampere equation:

$$\left(\frac{\partial^2 w}{\partial x \partial y}\right)^2 - \frac{\partial^2 w}{\partial x^2}\frac{\partial^2 w}{\partial y^2} = f(x, y).$$

The equation is encountered in differential geometry, gas dynamics, and meteorology.

Below are solutions to the *homogeneous Monge–Ampere equation* for the special case $f(x, y) \equiv 0$.

1. Exact solutions involving one arbitrary function:

$$w(x, y) = \varphi(C_1 x + C_2 y) + C_3 x + C_4 y + C_5,$$
$$w(x, y) = (C_1 x + C_2 y)\varphi(yx) + C_3 x + C_4 y + C_5,$$
$$w(x, y) = (C_1 x + C_2 y + C_3)\varphi\left(\frac{C_4 x + C_5 y + C_6}{C_1 x + C_2 y + C_3}\right) + C_7 x + C_8 y + C_9,$$

where $C_1|, \ldots, C_9$ are arbitrary constants and $\varphi = \varphi(z)$ is an arbitrary function.

2. General solution in parametric form:

$$w = tx + \varphi(t)y + \psi(t),$$

$$x + \varphi'(t)y + \psi'(t) = 0,$$

Where t is the parameter, and $\varphi=\varphi(t)$ and $\psi=\psi(t)$ are arbitrary functions.

Simplest Types of Exact Solutions of Nonlinear PDEs

The following classes of solutions are usually regarded as exact solutions to nonlinear partial differential equations of mathematical physics:

1. Solutions expressible in terms of elementary functions.

2. Solutions expressed by quadrature.

3. Solutions described by ordinary differential equations (or systems of ordinary differential equations).

4. Solutions expressible in terms of solutions to linear partial differential equations (and/or solutions to linear integral equations).

The simplest types of exact solutions to nonlinear PDEs are traveling-wave solutions and self-similar solutions. They often occur in various applications.

In what follows, it is assumed that the unknown w depends on two variables, x and t, where t plays the role of time and x is a spatial coordinate.

Traveling-wave Solutions

Traveling-wave solutions, by definition, are of the form:

$$w(x,t) = W(z), \quad z = kx - \lambda t,$$

where λ/k plays the role of the wave propagation velocity (the value $\lambda=0$ corresponds to a stationary solution, and the value $k=0$ corresponds to a space-homogeneous solution). Traveling-wave solutions are characterized by the fact that the profiles of these solutions at different time instants are obtained from one another by appropriate shifts (translations) along the x-axis. Consequently, a Cartesian coordinate system moving with a constant speed can be introduced in which the profile of the desired quantity is stationary. For $k>0$ and $\lambda>0$, the wave $w(x,t) = W(z), \quad z = kx - \lambda t$, travels along the x-axis to the right (in the direction of increasing x).

Traveling-wave solutions occur for equations that do not explicitly involve independent variables,

$$F\left(w, \frac{\partial w}{\partial x}, \frac{\partial w}{\partial t}, \frac{\partial^2 w}{\partial x^2}, \frac{\partial^2 w}{\partial x \partial t}, \frac{\partial^2 w}{\partial t_2}, \dots \right) = 0$$

Substituting $w(x,t) = W(z), \quad z = kx - \lambda t$, into $F\left(w, \frac{\partial w}{\partial x}, \frac{\partial w}{\partial t}, \frac{\partial^2 w}{\partial x^2}, \frac{\partial^2 w}{\partial x \partial t}, \frac{\partial^2 w}{\partial t_2}, \dots \right) = 0$, one obtains an autonomous ordinary differential equation for the function $W(z)$:

$$F(W, kW', -\lambda W', k^2 W'', -k\lambda W'', \lambda^2 W'', \dots) = 0,$$

Where $k|$ and λ are arbitrary constants, and the prime denotes a derivative with respect to z.

The term *traveling-wave solution* is also used in the cases where the variable t plays the role of a spatial coordinate, $t = y$.

All nonlinear equations considered above, and with $f(x,y)=0$, admit traveling-wave solutions.

Self-similar Solutions

By definition, a *self-similar solution* is a solution of the form:

$$w(x,t) = t^{\alpha} U(\zeta), \quad \zeta = xt^{\beta}.$$

The profiles of these solutions at different time instants are obtained from one another by a similarity transformation (like scaling).

Self-similar solutions exist if the scaling of the independent and dependent variables,

$t = C\overline{t}, \ x = C^k \overline{x}, \ w = C^m \overline{w}$, where $C \neq 0$ is an arbitrary constant,

for some k and m such that $|k| + |m| \neq 0$, is equivalent to the identical transformation.

It can be shown that the parameters in solution $w(x,t) = t^{\alpha} U(\zeta), \quad \zeta = xt^{\beta}$ and transformation $t = C\overline{t}, \ x = C^k \overline{x}, \ w = C^m \overline{w}$, are linked by the simple relations:

$$\alpha = m, \quad \beta = -k.$$

In practice, the above existence criterion is checked and if a pair of k and m in has been found, then a self-similar solution is defined by formulas $w(x,t) = t^{\alpha} U(\zeta), \quad \zeta = xt^{\beta}$ with parameters $\alpha = m, \quad \beta = -k$.

Example: Consider the heat equation with a nonlinear power-law source term-

$$\frac{\partial w}{\partial t} = a \frac{\partial^2 w}{\partial x^2} + bw^n.$$

The scaling transformation $t = C\overline{t}, \ x = C^k \overline{x}, \ w = C^m \overline{w}$, converts equation $\dfrac{\partial w}{\partial t} = a \dfrac{\partial^2 w}{\partial x^2} + bw^n$ into

$$C^{m-1} \frac{\partial \overline{w}}{\partial \overline{t}} = aC^{m-2k} \frac{\partial^2 \overline{w}}{\partial \overline{x}^2} + bC^{mn} \overline{w}^n.$$

In order that equation $C^{m-1} \dfrac{\partial \overline{w}}{\partial \overline{t}} = aC^{m-2k} \dfrac{\partial^2 \overline{w}}{\partial \overline{x}^2} + bC^{mn} \overline{w}^n$ coincides with $\dfrac{\partial w}{\partial t} = a \dfrac{\partial^2 w}{\partial x^2} + bw^n$, one must require that the powers of C are the same, which yields the following system of linear algebraic equations for the constants k and m:

$$m - 1 = m - 2k = mn.$$

This system admits a unique solution:

$$m = \frac{1}{1-n}.$$

Using this solution together with relations $w(x,t) = t^\alpha U(\zeta)$, $\zeta = xt^\beta$ and $\dfrac{\partial w}{\partial t} = a\dfrac{\partial^2 w}{\partial x^2} + bw^n$, one obtains self-similar variables in the form:

$$w = t^{1/(1-n)}U(\zeta), \qquad \zeta = xt^{-1/2}.$$

Inserting these into, one arrives at the following ordinary differential equation for $U(\zeta)$:

$$aU''_{\zeta\zeta} + \frac{1}{2}\zeta U'_\zeta + \frac{1}{n-1}U + bU^n = 0.$$

Cauchy Problem and Boundary Value Problems for Nonlinear Equations

The Cauchy problem and boundary value problems for nonlinear equations are stated in exactly the same way as for linear equations.

Examples: The Cauchy problem for a nonlinear heat equation is stated as follows.

Find a solution to equation $\dfrac{\partial w}{\partial t} = \dfrac{\partial}{\partial x}\left[f(w)\dfrac{\partial w}{\partial x}\right]$. subject to the initial condition.

The first boundary value problem for a nonlinear wave equation as follows: find a solution to equation $\dfrac{\partial^2 w}{\partial x^2} + \dfrac{\partial^2 w}{\partial y^2} = f(w)$. subject to the initial conditions and the boundary conditions.

Problems for nonlinear PDEs are normally solved using numerical methods.

Matrix

A rectangular array of numbers, symbols and expressions arranged in rows and columns is known as a matrix. Some of the focus areas of matrix are determinant, invertible matrix, Cayley-Hamilton theorem, LU decompositions, Eigen values and vectors, etc. These diverse areas of matrix have been thoroughly discussed in this chapter.

A rectangular array of m x n numbers (real or complex) in the form of m horizontal lines (called rows) and n vertical lines (called columns), is called a matrix of order m by n, written as m x n matrix. Such an array is enclosed by [ ] or ( ).

An m x n matrix is usually written as:

$$A = \begin{bmatrix} a_{11} & a_{12} & \ldots & a_{1n} \\ a_{21} & a_{22} & \ldots & a_{2n} \\ \vdots & \vdots & \vdots & \vdots \\ a_{m1} & a_{m2} & \ldots & a_{mn} \end{bmatrix}$$

In brief, the above matrix is represented by $A = [a_{ij}]_{mxn}$. The number a_{11}, a_{12}, ... etc., are known as the elements of the matrix A, where a_{ij} belongs to the ith row and j^{th} column and is called the $(i, j)^{th}$ element of the matrix $A = [a_{ij}]$.

Important Formulas for Matrices

If A, B are square matrices of order n, and I_n is a corresponding unit matrix, then:

$$A(adj.A) = |A| I_n = (adj\ A)\ A$$

$$|adj\ A| = |A|n^{-1} \left(\text{Thus } A\ (adj\ A) \text{ is always a scalar matrix}\right)$$

$$adj\ (adj.A) = |A|^{n-2}\ A$$

$$|adj(adj.A)| = |A|^{(n-1)}\ 2$$

$$adj\ (AB) = (adj\ B)\ (adj\ A)$$

$$adj\ (A^m) = (adj\ A)^m,$$

$$adj(kA) = k^{n-1}(adj.A), k \in R$$

$$adj(I_n) = I_n$$

$$adj\ o = o$$

A is symmetric ⇒adj A is also symmetric

A is diagonal ⇒adj A is also diagonal

A is triangular ⇒adj A is also triangular

A is singular ⇒| adj A | = o

Types of Matrices

1. Symmetric Matrix: A square matrix A =$[a_{ij}]$ is called a symmetric matrix if $a_{ij}=a_{ji}$, for all i, j.

2. Skew-Symmetric Matrix: when $a_{ij}=-a_{ji}$.

3. Hermitian and skew – Hermitian Matrix: A=A^θ (Hermitian matrix).

   $A^\theta=-A$ (skew-Hermitian matrix).

4. Orthogonal matrix: if $AA^T = I_n = A^TA$.

5. Idempotent matrix: if $A^2 = A$.

6. Involuntary matrix: if $A^2 = I$ or $A^{-1} = A$.

7. Nilpotent matrix: if ∃p∈N such that A^p=o.

Trace of Matrix

$$tr(\lambda A) = \lambda tr(A)$$
$$tr(A+B) = tr(A) + tr(B)$$
$$tr(AB) = tr(BA)$$

Transpose of Matrix

$$(A^T)^T = A$$
$$(A \pm B)^T = A^T \pm B^T$$
$$(AB)^T = B^T A^T$$
$$(kA)^T = k(A)^T$$
$$(A_1 A_2 A_3 \ldots\ldots A_{n-1} A_n)^T = A_n^T A_{n-1}^t \ldots\ldots A_3^T A_2^T$$
$$I^T = I$$
$$tr(A) = t(A^T)$$

Properties of Matrix Multiplication

$$AB \neq BA$$
$$(AB)C = A(BC)$$
$$A.(B+C) = A.B + A.C$$

Adjoint of a Matrix

$$A(adj\ A) = (adj\ A)\ A = |\ A\ |\ I_n$$

$$adjA\ |=|\ A\ |^{n-1}$$

$$(adj\ AB) = (adj\ B)(adj\ A)$$

$$adj(adjA) =|\ A\ |^{n-2}$$

Inverse of a Matrix

A^{-1} exists if A is non singular i.e. $|A| \neq 0$

$$A^{-1} = \frac{1}{|A|}(Adj.A)$$

$$A^{-1}A = I_n = AA^{-1}$$

$$(A^T)^{-1} = (A^{-1})^T$$

$$(A^{-1})^{-1} = A$$

$$|A^{-1}| = |A|^{-1} = \frac{1}{|A|}$$

Order of a Matrix

A matrix which has m rows and n columns is called a matrix of order m x n.

E.g. the order of $\begin{bmatrix} 4 & -1 & 5 \\ 6 & 8 & -7 \end{bmatrix}$ matrix is 2 x 3.

Note: (a) The matrix is just an arrangement of certain quantities.

(b) The elements of a matrix may be real or complex numbers. If all the elements of a matrix are real, then the matrix is called a real matrix.

(c) An m x n matrix has m.n elements.

Illustration: Construct a 3×4 matrix A = $[a_{ij}]$, whose elements are given by $a_{ij} = 2i + 3j$.

$$\begin{bmatrix} a_{11} & a_{12} & a_{13} & a_{14} \\ a_{21} & a_{22} & a_{23} & a_{24} \\ a_{31} & a_{32} & a_{33} & a_{34} \end{bmatrix} ; \quad \therefore a_{11} = 2 \times 1 + 3 \times 1 = 5; a_{12} = 2 \times 1 + 3 \times 2 = 8.$$

Solution: In this problem, I and j are the number of rows and columns respectively. By substituting the respective values of rows and columns in $a_{ij} = 2i + 3j$ we can construct the required matrix.

Similarly, a_{13} = 11, a_{14}=14, a_{21} = 7, a_{22}=10, a_{23}=13, a_{24}=16, a_{31}=9, a_{32}=12, a_{33}=15, a_{34}=18

$$\therefore A = \begin{bmatrix} 5 & 8 & 11 & 14 \\ 7 & 10 & 13 & 16 \\ 9 & 12 & 18 & 18 \end{bmatrix}$$

Illustration: Construct a 3 x 4 matrix, whose elements are given by: $a_{ij} = \dfrac{1}{2}|3i+j|$.

Solution: Method for solving this problem is the same as in the above problem.

Since $a_{ij} = \dfrac{1}{2}|-3i+j|$ *we have* $a_{11} = \dfrac{1}{2}|-3(1)+1| = \dfrac{1}{2}|-3+1| = \dfrac{1}{2}|-2| = \dfrac{2}{2} = 1$

$a_{12} = \dfrac{1}{2}|-3(1)+2| = \dfrac{1}{2}|-3+2| = \dfrac{1}{2}|-1| = \dfrac{1}{2}$

$a_{13} = \dfrac{1}{2}|-3(1)+3| = \dfrac{1}{2}|-3+3| = \dfrac{1}{2}(0) = 0$

$a_{14} = \dfrac{1}{2}|-3(1)+4| = \dfrac{1}{2}|-3+4| = \dfrac{1}{2}; \qquad a_{21}\dfrac{1}{2}|-3(2)+1| = \dfrac{1}{2}|-6+1| = \dfrac{5}{2}$

$a_{22} = \dfrac{1}{2}|-3(2)+2| = \dfrac{1}{2}|-6+2| = \dfrac{4}{2} = 2; \qquad a_{23}\dfrac{1}{2}|-3(2)+3| = \dfrac{1}{2}|-6+3| = \dfrac{3}{2}$

$a_{24} = \dfrac{1}{2}|-3(2)+4| = \dfrac{1}{2}|-6+4| = \dfrac{2}{2} = 1; \qquad$ *Similarly* $a_{31}=4, a_{32}=\dfrac{7}{2}, a_{33}=3, a_{34}=\dfrac{5}{2}$

Hence, the required matrix is given by $A = \begin{bmatrix} 1 & \dfrac{1}{2} & 0 & \dfrac{1}{2} \\ \dfrac{5}{2} & 2 & \dfrac{3}{2} & 1 \\ 4 & \dfrac{7}{2} & 3 & \dfrac{5}{2} \end{bmatrix}$.

Trace of a Matrix

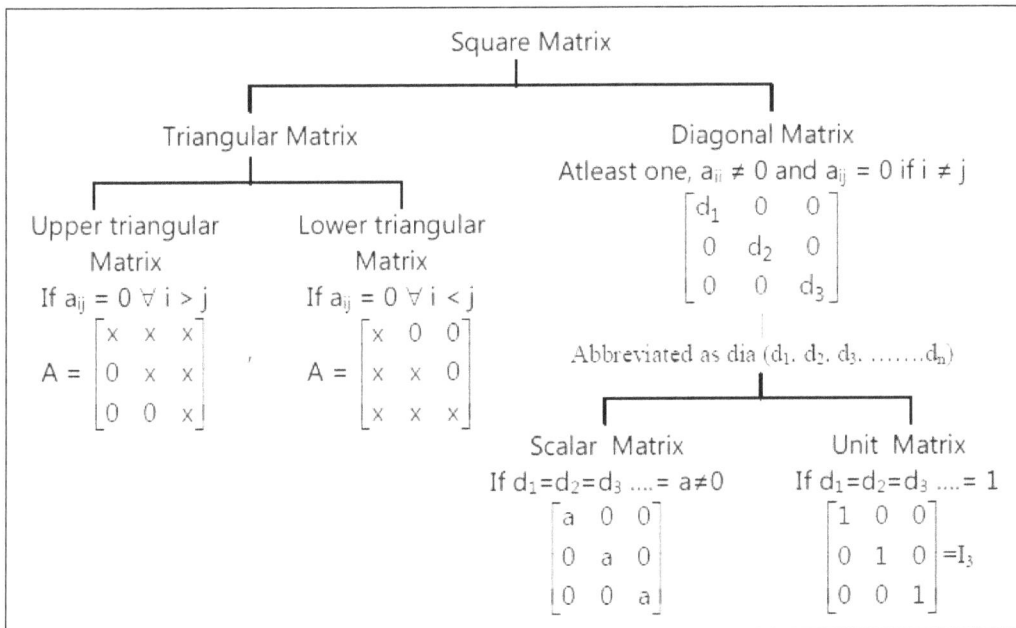

Let A = $[a_{ij}]_{nxn}$ and B = $[b_{ij}]_{nxn}$ and λ be a scalar,

(i) tr(λA) = λ tr(A) (ii) tr(A + B) = tr(A) + tr(B) (iii) tr(AB) = tr(BA)

Transpose of Matrix

The matrix obtained from a given matrix A by changing its rows into columns or columns into rows is called the transpose of matrix A and is denoted by A^T or A'. From the definition it is obvious that if the order of A is m x n, then the order of A^T becomes n x m; E.g. transpose of matrix:

$$\begin{bmatrix} a1 & a2 & a3 \\ b1 & b2 & b3 \end{bmatrix}_{2\times3} \quad is \begin{bmatrix} a1 & b1 \\ a2 & b2 \\ a3 & b3 \end{bmatrix}_{3\times2}$$

Properties of Transpose of Matrix

$$\left(A^T\right)^T = A$$

$$\left(A+B\right)^T = A^T + B^T$$

$$\left(AB\right)^T = B^T A^T$$

$$\left(kA\right)^T = k\left(A\right)^T$$

$$\left(A_1 A_2 A_3 \,......A_{n-1} A_n\right)^T = A_n^T A_{n-1}^T .....A_3^T A_2^T A_1^T$$

$$I^T = I$$

$$tr\left(A\right) = tr\left(A^T\right)$$

DETERMINANT OF A MATRIX

The determinant of a matrix is a number that is specially defined only for square matrices. Determinants are mathematical objects that are very useful in the analysis and solution of systems of linear equations. Determinants also have wide applications in engineering, science, economics and social science as well.

To every square matrix A = [aij] of order n, we can associate a number (real or complex) called determinant of the square matrix A, where a = (i, j)th element of A. This may be thought of as a function which associates each square matrix with a unique number (real or complex).

If M is the set of square matrices, K is the set of numbers (real or complex) and f : M → K is defined by f (A) = k, where A ∈ M and k ∈ K, then f (A) is called the determinant of A. It is also denoted by | A | or det A or Δ.

$$If \; A = \begin{bmatrix} a & b \\ c & d \end{bmatrix}, then \, determinant \, of \, A \, is \, written \, as \; |A| = \begin{bmatrix} a & b \\ c & d \end{bmatrix} = det \, A$$

For a 1×1 Matrix

Let A = [a] be the matrix of order 1, then determinant of A is defined to be equal to a.

For a 2×2 Matrix

For a 2×2 matrix (2 rows and 2 columns):

$$A = \begin{bmatrix} a & b \\ c & d \end{bmatrix}$$

The determinant is: |A| = ad − bc or the determinant of A equals a × d minus b × c. It is easy to remember when you think of a cross, where blue is positive that goes diagonally from left to right and red is negative that goes diagonally from right to left.

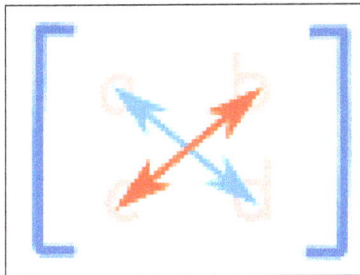

Example:

$$If\ A = \begin{bmatrix} 2 & 3 \\ 4 & 8 \end{bmatrix}$$

|A| = 2 x 8 − 4 x 3
= 16 − 12
= 4

For a 3×3 Matrix

For a 3×3 matrix (3 rows and 3 columns):

$$A = \begin{bmatrix} a & b & c \\ d & e & f \\ g & h & i \end{bmatrix}$$

The determinant is: |A| = a (ei − fh) − b (di − fg) + c (dh − eg). The determinant of A equals 'a times e x i minus f x h minus b times d x i minus f x g plus c times d x h minus e x g'. It may look complicated, but if you carefully observe the pattern its really easy.

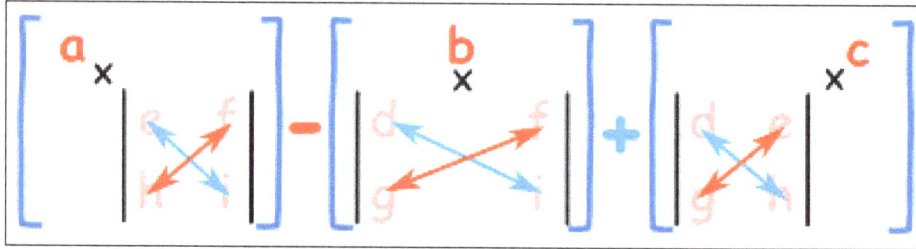

To work out the determinant of a matrix 3×3:

- Multiply 'a' by the determinant of the 2×2 matrix that is not in a's row or column.

- Likewise for 'b' and for 'c'.

- Sum them up, but remember the minus in front of the b.

As a formula (remember the vertical bars || mean "determinant of"):

$$|A| = a.\begin{vmatrix} e & f \\ h & i \end{vmatrix} - b.\begin{vmatrix} d & f \\ g & i \end{vmatrix} + c.\begin{vmatrix} d & e \\ g & h \end{vmatrix}$$

The determinant of A equals 'a' times the determinant of e × i minus f × h minus 'b' times the determinant of d × i minus f × g plus 'c' times the determinant of d × h minus e × g.

Example:

$$If\ A = \begin{bmatrix} 6 & 1 & 1 \\ 4 & -2 & 5 \\ 2 & 8 & 7 \end{bmatrix}$$

$$|A| = 6 \times (-2 \times 7 - 5 \times 8) - 1 \times (4 \times 7 - 5 \times 2) + 1 \times (4 \times 8 - (-2 \times 2))$$

$$= 6 \times (-54) - 1 \times (18) + 1 \times (36)$$

$$= -306$$

For 4×4 Matrices and Higher

The pattern continues for the determinant of a matrix 4×4:

- Plus a times the determinant of the matrix that is not in a's row or column.

- Minus b times the determinant of the matrix that is not in b's row or column.

- Plus c times the determinant of the matrix that is not in c's row or column.

- Minus d times the determinant of the matrix that is not in d's row or column.

As a formula:

$$|A| = a.\begin{vmatrix} f & g & h \\ j & k & l \\ n & o & p \end{vmatrix} - b.\begin{vmatrix} e & g & h \\ i & k & l \\ m & o & p \end{vmatrix} + c.\begin{vmatrix} e & f & h \\ i & j & l \\ m & n & p \end{vmatrix} - d.\begin{vmatrix} e & f & g \\ i & j & k \\ m & n & o \end{vmatrix}$$

Notice the +−+− pattern (+a... −b... +c... −d...).

INVERTIBLE MATRIX

In linear algebra, an n-by-n square matrix A is called invertible (also nonsingular or nondegenerate) if there exists an n-by-n square matrix B such that:

$$AB = BA = I_n$$

where I_n denotes the n-by-n identity matrix and the multiplication used is ordinary matrix multiplication. If this is the case, then the matrix B is uniquely determined by A and is called the *inverse* of A, denoted by A^{-1}.

A square matrix that is not invertible is called singular or degenerate. A square matrix is singular if and only if its determinant is 0. Singular matrices are rare in the sense that a square matrix randomly selected from a continuous uniform distribution on its entries will almost never be singular.

Non-square matrices (m-by-n matrices for which $m \neq n$) do not have an inverse. However, in some cases such a matrix may have a left inverse or right inverse. If A is m-by-n and the rank of A is equal to n ($n \leq m$), then A has a left inverse: an n-by-m matrix B such that $BA = I_n$. If A has rank m ($m \leq n$), then it has a right inverse: an n-by-m matrix B such that $AB = I_m$.

Matrix inversion is the process of finding the matrix B that satisfies the prior equation for a given invertible matrix A.

While the most common case is that of matrices over the real or complex numbers, all these definitions can be given for matrices over any ring. However, in the case of the ring being commutative, the condition for a square matrix to be invertible is that its determinant is invertible in the ring, which in general is a stricter requirement than being nonzero. For a noncommutative ring, the usual determinant is not defined. The conditions for existence of left-inverse or right-inverse are more complicated since a notion of rank does not exist over rings.

The set of $n \times n$ invertible matrices together with the operation of matrix multiplication form a group, the general linear group of degree n.

Properties

Invertible Matrix Theorem

Let A be a square n by n matrix over a field K (for example the field R of real numbers). The following statements are equivalent, that is, for any given matrix they are either all true or all false:

- A is invertible, that is, A has an inverse, is nonsingular, or is nondegenerate.

- A is row-equivalent to the n-by-n identity matrix I_n.

- A is column-equivalent to the n-by-n identity matrix I_n.

- A has n pivot positions.

- det A $\neq$ 0. In general, a square matrix over a commutative ring is invertible if and only if its determinant is a unit in that ring.

- A has full rank; that is, rank A = n.

- The equation Ax = 0 has only the trivial solution x = 0.

- The kernel of A is trivial, that is, it contains only the null vector as an element, $ker(A) = \{0\}$.

- Null A = {0}.

- The equation Ax = b has exactly one solution for each b in K^n.

- The columns of A are linearly independent.

- The columns of A span K^n.

- Col A = K^n.

- The columns of A form a basis of K^n.

- The linear transformation mapping x to Ax is a bijection from K^n to K^n.

- There is an n-by-n matrix B such that $AB = I_n = BA$.

- The transpose A^T is an invertible matrix (hence rows of A are linearly independent, span K^n, and form a basis of K^n).

- The number 0 is not an eigenvalue of A.

- The matrix A can be expressed as a finite product of elementary matrices.

- The matrix A has a left inverse (that is, there exists a B such that BA = I) *or* a right inverse (that is, there exists a C such that AC = I), in which case both left and right inverses exist and $B = C = A^{-1}$.

Other Properties

Furthermore, the following properties hold for an invertible matrix **A**:

- $(A^{-1})^{-1} = A$;

- $(kA)^{-1} = k^{-1}A^{-1}$ for nonzero scalar k;

- $(Ax)^+ = x^+A^{-1}$ where $^+$ denotes the Moore–Penrose inverse and x is a vector;

- $(A^T)^{-1} = (A^{-1})^T$;

- For any invertible n-by-n matrices A and B, $(AB)^{-1} = B^{-1}A^{-1}$. More generally, if $A_1,\dots,A_k$ are invertible n-by-n matrices, then $(A_1A_2\cdots A_{k-1}A_k)^{-1} = A_k^{-1}A_{k-1}^{-1}\cdots A_2^{-1}A_1^{-1}$;

- $\det A^{-1} = (\det A)^{-1}$.

The rows of the inverse matrix V of a matrix U are orthonormal to the columns of U (and vice versa interchanging rows for columns). To see this, suppose that UV = VU = I where we write the rows of V as v_i^T and the columns of U as u_j for $1 \leq i, j \leq n$.. Then clearly, the Euclidean inner product of any two $v_i^T u_j = \delta_{i,j}$. This property can also be useful in constructing the inverse of a square matrix in some instances where a set of orthogonal vectors (but not necessarily orthonormal vectors) to the columns of U are known and then applying the iterative Gram–Schmidt process to this initial set to determine the rows of the inverse V.

A matrix that is its own inverse, that is, such that $A = A^{-1}$ and $A^2 = I$, is called an involutory matrix.

In relation to its adjugate, the adjugate of a matrix A can be used to find the inverse of A as follows:

If A is an $n \times n$ invertible matrix, then:

$$A^{-1} = \frac{1}{\det(A)} \operatorname{adj}(A).$$

In relation to the identity matrix:

It follows from the associativity of matrix multiplication that if:

$$AB = I$$

for *finite square* matrices A and B, then also:

$$BA = I$$

Density

Over the field of real numbers, the set of singular n-by-n matrices, considered as a subset of $R^{n \times n}$, is a null set, that is, has Lebesgue measure zero. This is true because singular matrices are the roots of the determinant function. This is a continuous function because it is a polynomial in the entries of the matrix. Thus in the language of measure theory, almost all n-by-n matrices are invertible.

Furthermore, the *n*-by-*n* invertible matrices are a dense open set in the topological space of all *n*-by-*n* matrices. Equivalently, the set of singular matrices is closed and nowhere dense in the space of *n*-by-*n* matrices.

In practice however, one may encounter non-invertible matrices. And in numerical calculations, matrices which are invertible, but close to a non-invertible matrix, can still be problematic; such matrices are said to be ill-conditioned.

Consider the following 2-by-2 matrix:

$$\mathbf{A} = \begin{pmatrix} -1 & \frac{3}{2} \\ 1 & -1 \end{pmatrix}$$

The matrix $\mathbf{A}$ is invertible. To check this, one can compute that $\det \mathbf{A} = -1/2$, which is non-zero.

As an example of a non-invertible, or singular, matrix, consider the matrix:

$$\mathbf{B} = \begin{pmatrix} -1 & \frac{3}{2} \\ \frac{2}{3} & -1 \end{pmatrix}.$$

The determinant of $\mathbf{B}$ is 0, which is a necessary and sufficient condition for a matrix to be non-invertible.

Methods of Matrix Inversion

Gaussian Elimination

Gauss–Jordan elimination is an algorithm that can be used to determine whether a given matrix is invertible and to find the inverse. An alternative is the LU decomposition, which generates upper and lower triangular matrices, which are easier to invert.

Newton's Method

A generalization of Newton's method as used for a multiplicative inverse algorithm may be convenient, if it is convenient to find a suitable starting seed:

$$X_{k+1} = 2X_k - X_k A X_k.$$

Victor Pan and John Reif have done work that includes ways of generating a starting seed. Byte magazine summarised one of their approaches.

Newton's method is particularly useful when dealing with families of related matrices that behave enough like the sequence manufactured for the homotopy above: sometimes a good starting point for refining an approximation for the new inverse can be the already obtained inverse of a previous matrix that nearly matches the current matrix, for example, the pair of sequences of inverse matrices used in obtaining matrix square roots by Denman–Beavers iteration; this may need more than one pass of the iteration at each new matrix, if they are not close enough together for just one to be enough. Newton's method is also useful for "touch up" corrections to the Gauss–Jordan algorithm which has been contaminated by small errors due to imperfect computer arithmetic.

Cayley–Hamilton Method

The Cayley–Hamilton theorem allows the inverse of **A** to be expressed in terms of det(**A**), traces and powers of **A**:

$$\mathbf{A}^{-1} = \frac{1}{\det(\mathbf{A})} \sum_{s=0}^{n-1} \mathbf{A}^s \sum_{k_1, k_2, \ldots, k_{n-1}} \prod_{l=1}^{n-1} \frac{(-1)^{k_l+1}}{l^{k_l} k_l!} \operatorname{tr}(\mathbf{A}^l)^{k_l},$$

where n is dimension of **A**, and $\operatorname{tr}(A)$ is the trace of matrix A given by the sum of the main diagonal. The sum is taken over s and the sets of all $k_l \geq 0$ satisfying the linear Diophantine equation:

$$s + \sum_{l=1}^{n-1} l k_l = n - 1.$$

The formula can be rewritten in terms of complete Bell polynomials of arguments $t_l = -(l-1)! \operatorname{tr}(A^l)$ as:

$$\mathbf{A}^{-1} = \frac{1}{\det(\mathbf{A})} \sum_{s=1}^{n} \mathbf{A}^{s-1} \frac{(-1)^{n-1}}{(n-s)!} B_{n-s}(t_1, t_2, \ldots, t_{n-s}).$$

Eigendecomposition

If matrix A can be eigendecomposed, and if none of its eigenvalues are zero, then A is invertible and its inverse is given by:

$$\mathbf{A}^{-1} = \mathbf{Q}\Lambda^{-1}\mathbf{Q}^{-1},$$

where Q is the square (*N*×*N*) matrix whose *i*-th column is the eigenvector q_i of A, and Λ is the diagonal matrix whose diagonal elements are the corresponding eigenvalues, that is, $\Lambda_{ii} = \lambda_i$. Furthermore, because Λ is a diagonal matrix, its inverse is easy to calculate:

$$\left[\Lambda^{-1}\right]_{ii} = \frac{1}{\lambda_i}.$$

Cholesky Decomposition

If matrix **A** is positive definite, then its inverse can be obtained as:

$$\mathbf{A}^{-1} = (\mathbf{L}^*)^{-1}\mathbf{L}^{-1},$$

where L is the lower triangular Cholesky decomposition of A, and L* denotes the conjugate transpose of L.

Analytic Solution

Writing the transpose of the matrix of cofactors, known as an adjugate matrix, can also be an efficient way to calculate the inverse of *small* matrices, but this recursive method is inefficient for large matrices. To determine the inverse, we calculate a matrix of cofactors:

$$\mathbf{A}^{-1} = \frac{1}{|\mathbf{A}|}\mathbf{C}^{\mathrm{T}} = \frac{1}{|\mathbf{A}|}\begin{pmatrix} \mathbf{C}_{11} & \mathbf{C}_{21} & \cdots & \mathbf{C}_{n1} \\ \mathbf{C}_{12} & \mathbf{C}_{22} & \cdots & \mathbf{C}_{n2} \\ \vdots & \vdots & \ddots & \vdots \\ \mathbf{C}_{1n} & \mathbf{C}_{2n} & \cdots & \mathbf{C}_{nn} \end{pmatrix}$$

so that,

$$\left(\mathbf{A}^{-1}\right)_{ij} = \frac{1}{|\mathbf{A}|}\left(\mathbf{C}^{\mathrm{T}}\right)_{ij} = \frac{1}{|\mathbf{A}|}\left(\mathbf{C}_{ji}\right)$$

where $|\mathbf{A}|$ is the determinant of A, C is the matrix of cofactors, and $\mathbf{C}^{\mathrm{T}}$ represents the matrix transpose.

Inversion of 2 × 2 matrices

The *cofactor equation* listed above yields the following result for 2 × 2 matrices. Inversion of these matrices can be done as follows:

$$\mathbf{A}^{-1} = \begin{bmatrix} a & b \\ c & d \end{bmatrix}^{-1} = \frac{1}{\det \mathbf{A}}\begin{bmatrix} d & -b \\ -c & a \end{bmatrix} = \frac{1}{ad-bc}\begin{bmatrix} d & -b \\ -c & a \end{bmatrix}.$$

This is possible because $1/(ad - bc)$ is the reciprocal of the determinant of the matrix in question, and the same strategy could be used for other matrix sizes.

The Cayley–Hamilton method gives:

$$\mathbf{A}^{-1} = \frac{1}{\det \mathbf{A}}\left[(\operatorname{tr}\mathbf{A})\mathbf{I} - \mathbf{A}\right].$$

Inversion of 3 × 3 matrices

A computationally efficient 3 × 3 matrix inversion is given by:

$$\mathbf{A}^{-1} = \begin{bmatrix} a & b & c \\ d & e & f \\ g & h & i \end{bmatrix}^{-1} = \frac{1}{\det(\mathbf{A})}\begin{bmatrix} A & B & C \\ D & E & F \\ G & H & I \end{bmatrix}^{\mathrm{T}} = \frac{1}{\det(\mathbf{A})}\begin{bmatrix} A & D & G \\ B & E & H \\ C & F & I \end{bmatrix}$$

(where the scalar A is the matrix A). If the determinant is non-zero, the matrix is invertible, with the elements of the intermediary matrix on the right side above given by:

$$A = (ei - fh), \quad D = -(bi - ch), \quad G = (bf - ce),$$

$$B = -(di - fg), \quad E = (ai - cg), \quad H = -(af - cd),$$
$$C = (dh - eg), \quad F = -(ah - bg), \quad I = (ae - bd).$$

The determinant of A can be computed by applying the rule of Sarrus as follows:

$$\det(\mathbf{A}) = aA + bB + cC.$$

The Cayley–Hamilton decomposition gives:

$$\mathbf{A}^{-1} = \frac{1}{\det(\mathbf{A})} \left[\frac{1}{2} \left((\operatorname{tr} \mathbf{A})^2 - \operatorname{tr} \mathbf{A}^2 \right) \mathbf{I} - \mathbf{A} \operatorname{tr} \mathbf{A} + \mathbf{A}^2 \right].$$

The general 3 × 3 inverse can be expressed concisely in terms of the cross product and triple product. If a matrix $\mathbf{A} = [\mathbf{x}_0, \mathbf{x}_1, \mathbf{x}_2]$ (consisting of three column vectors, $\mathbf{x}_0$, $\mathbf{x}_1$, and $\mathbf{x}_2$) is invertible, its inverse is given by:

$$\mathbf{A}^{-1} = \frac{1}{\det(\mathbf{A})} \begin{bmatrix} (\mathbf{x}_1 \times \mathbf{x}_2)^{\mathrm{T}} \\ (\mathbf{x}_2 \times \mathbf{x}_0)^{\mathrm{T}} \\ (\mathbf{x}_0 \times \mathbf{x}_1)^{\mathrm{T}} \end{bmatrix}.$$

Note that $\det(\mathbf{A})$ is equal to the triple product of $\mathbf{x}_0$, $\mathbf{x}_1$, and $\mathbf{x}_2$ —the volume of the parallelepiped formed by the rows or columns:

$$\det(\mathbf{A}) = \mathbf{x}_0 \cdot (\mathbf{x}_1 \times \mathbf{x}_2).$$

The correctness of the formula can be checked by using cross- and triple-product properties and by noting that for groups, left and right inverses always coincide. Intuitively, because of the cross products, each row of $\mathbf{A}^{-1}$ is orthogonal to the non-corresponding two columns of   (causing the off-diagonal terms of $\mathbf{I} = \mathbf{A}^{-1}\mathbf{A}$ be zero). Dividing by:

$$\det(\mathbf{A}) = \mathbf{x}_0 \cdot (\mathbf{x}_1 \times \mathbf{x}_2)$$

causes the diagonal elements of $\mathbf{I} = \mathbf{A}^{-1}\mathbf{A}$ to be unity. For example, the first diagonal is:

$$1 = \frac{1}{\mathbf{x}_0 \cdot (\mathbf{x}_1 \times \mathbf{x}_2)} \mathbf{x}_0 \cdot (\mathbf{x}_1 \times \mathbf{x}_2).$$

Inversion of 4 × 4 matrices

With increasing dimension, expressions for the inverse of $\mathbf{A}$ get complicated. For $n = 4$, the Cayley–Hamilton method leads to an expression that is still tractable:

$$\mathbf{A}^{-1} = \frac{1}{\det(\mathbf{A})} \left[\frac{1}{6} \left((\operatorname{tr} \mathbf{A})^3 - 3 \operatorname{tr} \mathbf{A} \operatorname{tr} \mathbf{A}^2 + 2 \operatorname{tr} \mathbf{A}^3 \right) \mathbf{I} - \frac{1}{2} \mathbf{A} \left((\operatorname{tr} \mathbf{A})^2 - \operatorname{tr} \mathbf{A}^2 \right) + \mathbf{A}^2 \operatorname{tr} \mathbf{A} - \mathbf{A}^3 \right].$$

Blockwise Inversion

Matrices can also be *inverted blockwise* by using the following analytic inversion formula:

$$\begin{bmatrix} \mathbf{A} & \mathbf{B} \\ \mathbf{C} & \mathbf{D} \end{bmatrix}^{-1} = \begin{bmatrix} \mathbf{A}^{-1} + \mathbf{A}^{-1}\mathbf{B}(\mathbf{D}-\mathbf{C}\mathbf{A}^{-1}\mathbf{B})^{-1}\mathbf{C}\mathbf{A}^{-1} & -\mathbf{A}^{-1}\mathbf{B}(\mathbf{D}-\mathbf{C}\mathbf{A}^{-1}\mathbf{B})^{-1} \\ -(\mathbf{D}-\mathbf{C}\mathbf{A}^{-1}\mathbf{B})^{-1}\mathbf{C}\mathbf{A}^{-1} & (\mathbf{D}-\mathbf{C}\mathbf{A}^{-1}\mathbf{B})^{-1} \end{bmatrix},$$

where A, B, C and D are matrix sub-blocks of arbitrary size. (A must be square, so that it can be inverted. Furthermore, A and D – CA⁻¹B must be nonsingular.) This strategy is particularly advantageous if A is diagonal and D – CA⁻¹B (the Schur complement of A) is a small matrix, since they are the only matrices requiring inversion.

This technique was reinvented several times and is due to Hans Boltz, who used it for the inversion of geodetic matrices, and Tadeusz Banachiewicz, who generalized it and proved its correctness.

The nullity theorem says that the nullity of A equals the nullity of the sub-block in the lower right of the inverse matrix, and that the nullity of B equals the nullity of the sub-block in the upper right of the inverse matrix.

The inversion procedure that led to equation above performed matrix block operations that operated on C and D first. Instead, if A and B are operated on first, and provided D and A – BD⁻¹C are nonsingular, the result is:

$$\begin{bmatrix} \mathbf{A} & \mathbf{B} \\ \mathbf{C} & \mathbf{D} \end{bmatrix}^{-1} = \begin{bmatrix} (\mathbf{A}-\mathbf{B}\mathbf{D}^{-1}\mathbf{C})^{-1} & -(\mathbf{A}-\mathbf{B}\mathbf{D}^{-1}\mathbf{C})^{-1}\mathbf{B}\mathbf{D}^{-1} \\ -\mathbf{D}^{-1}\mathbf{C}(\mathbf{A}-\mathbf{B}\mathbf{D}^{-1}\mathbf{C})^{-1} & \mathbf{D}^{-1}+\mathbf{D}^{-1}\mathbf{C}(\mathbf{A}-\mathbf{B}\mathbf{D}^{-1}\mathbf{C})^{-1}\mathbf{B}\mathbf{D}^{-1} \end{bmatrix}.$$

Equating equations above leads to:

$$(\mathbf{A}-\mathbf{B}\mathbf{D}^{-1}\mathbf{C})^{-1} = \mathbf{A}^{-1}+\mathbf{A}^{-1}\mathbf{B}(\mathbf{D}-\mathbf{C}\mathbf{A}^{-1}\mathbf{B})^{-1}\mathbf{C}\mathbf{A}^{-1}$$
$$(\mathbf{A}-\mathbf{B}\mathbf{D}^{-1}\mathbf{C})^{-1}\mathbf{B}\mathbf{D}^{-1} = \mathbf{A}^{-1}\mathbf{B}(\mathbf{D}-\mathbf{C}\mathbf{A}^{-1}\mathbf{B})^{-1}$$
$$\mathbf{D}^{-1}\mathbf{C}(\mathbf{A}-\mathbf{B}\mathbf{D}^{-1}\mathbf{C})^{-1} = (\mathbf{D}-\mathbf{C}\mathbf{A}^{-1}\mathbf{B})^{-1}\mathbf{C}\mathbf{A}^{-1}$$
$$\mathbf{D}^{-1}+\mathbf{D}^{-1}\mathbf{C}(\mathbf{A}-\mathbf{B}\mathbf{D}^{-1}\mathbf{C})^{-1}\mathbf{B}\mathbf{D}^{-1} = (\mathbf{D}-\mathbf{C}\mathbf{A}^{-1}\mathbf{B})^{-1}$$

Since a blockwise inversion of an $n \times n$ matrix requires inversion of two half-sized matrices and 6 multiplications between two half-sized matrices, it can be shown that a divide and conquer algorithm that uses blockwise inversion to invert a matrix runs with the same time complexity as the matrix multiplication algorithm that is used internally. There exist matrix multiplication algorithms with a complexity of $O(n^{2.3727})$ operations, while the best proven lower bound is $\Omega(n^2 \log n)$.

This formula simplifies significantly when the upper right block matrix B is the zero matrix. This formulation is useful when the matrices A and D have relatively simple inverse formulas (or pseudo inverses in the case where the blocks are not all square. In this special case, the block matrix inversion formula stated in full generality above becomes:

$$\begin{bmatrix} \mathbf{A} & \mathbf{0} \\ \mathbf{C} & \mathbf{D} \end{bmatrix}^{-1} = \begin{bmatrix} \mathbf{A}^{-1} & \mathbf{0} \\ -\mathbf{D}^{-1}\mathbf{C}\mathbf{A}^{-1} & \mathbf{D}^{-1} \end{bmatrix}.$$

By Neumann Series

If a matrix A has the property that:

$$\lim_{n\to\infty}(\mathbf{I}-\mathbf{A})^n = 0$$

then A is nonsingular and its inverse may be expressed by a Neumann series:

$$\mathbf{A}^{-1} = \sum_{n=0}^{\infty}(\mathbf{I}-\mathbf{A})^n.$$

Truncating the sum results in an "approximate" inverse which may be useful as a preconditioner. Note that a truncated series can be accelerated exponentially by noting that the Neumann series is a geometric sum. As such, it satisfies:

$$\sum_{n=0}^{2^L-1}(\mathbf{I}-\mathbf{A})^n = \prod_{l=0}^{L-1}(\mathbf{I}+(\mathbf{I}-\mathbf{A})^{2^l}).$$

Therefore, only $2L-2$ matrix multiplications are needed to compute 2^L terms of the sum.

More generally, if $\mathbf{A}$ is "near" the invertible matrix $\mathbf{X}$ in the sense that:

$$\lim_{n\to\infty}(\mathbf{I}-\mathbf{X}^{-1}\mathbf{A})^n = 0 \text{ or } \lim_{n\to\infty}(\mathbf{I}-\mathbf{A}\mathbf{X}^{-1})^n = 0$$

then A is nonsingular and its inverse is:

$$\mathbf{A}^{-1} = \sum_{n=0}^{\infty}\left(\mathbf{X}^{-1}(\mathbf{X}-\mathbf{A})\right)^n\mathbf{X}^{-1}.$$

If it is also the case that $\mathbf{A}-\mathbf{X}$ has rank 1 then this simplifies to:

$$\mathbf{A}^{-1} = \mathbf{X}^{-1} - \frac{\mathbf{X}^{-1}(\mathbf{A}-\mathbf{X})\mathbf{X}^{-1}}{1+\text{tr}(\mathbf{X}^{-1}(\mathbf{A}-\mathbf{X}))}.$$

P-adic Approximation

If A is a matrix with integer or rational coefficients and we seek a solution in arbitrary-precision rationals, then a p-adic approximation method converges to an exact solution in $O(n^4 \log^2 n)$, assuming standard $O(n^3)$ matrix multiplication is used. The method relies on solving n linear systems via Dixon's method of p-adic approximation (each in $O(n^3 \log^2 n)$) and is available as such in software specialized in arbitrary-precision matrix operations, for example, in IML.

Derivative of the Matrix Inverse

Suppose that the invertible matrix $\mathbf{A}$ depends on a parameter t. Then the derivative of the inverse of $\mathbf{A}$ with respect to t is given by:

$$\frac{d\mathbf{A}^{-1}}{dt} = -\mathbf{A}^{-1}\frac{d\mathbf{A}}{dt}\mathbf{A}^{-1}.$$

To derive the above expression for the derivative of the inverse of $\mathbf{A}$, one can differentiate the definition of the matrix inverse $\mathbf{A}^{-1}\mathbf{A} = \mathbf{I}$ and then solve for the inverse of $\mathbf{A}$:

$$\frac{d\mathbf{A}^{-1}\mathbf{A}}{dt} = \frac{d\mathbf{A}^{-1}}{dt}\mathbf{A} + \mathbf{A}^{-1}\frac{d\mathbf{A}}{dt} = \frac{d\mathbf{I}}{dt} = \mathbf{0}.$$

Subtracting $\mathbf{A}^{-1}\dfrac{d\mathbf{A}}{dt}$ from both sides of the above and multiplying on the right by $\mathbf{A}^{-1}$ gives the correct expression for the derivative of the inverse:

$$\frac{d\mathbf{A}^{-1}}{dt} = -\mathbf{A}^{-1}\frac{d\mathbf{A}}{dt}\mathbf{A}^{-1}.$$

Similarly, if ϵ is a small number then,

$$\left(\mathbf{A} + \epsilon\mathbf{X}\right)^{-1} = \mathbf{A}^{-1} - \epsilon\mathbf{A}^{-1}\mathbf{X}\mathbf{A}^{-1} + \mathcal{O}(\epsilon^2).$$

More generally, if:

$$\frac{df(\mathbf{A})}{dt} = \sum_i g_i(\mathbf{A})\frac{d\mathbf{A}}{dt}h_i(\mathbf{A}),$$

then,

$$f(\mathbf{A} + \epsilon\mathbf{X}) = f(\mathbf{A}) + \epsilon\sum_i g_i(\mathbf{A})\mathbf{X}h_i(\mathbf{A}) + \mathcal{O}(\epsilon^2).$$

Given a positive integer n,

$$\frac{d\mathbf{A}^n}{dt} = \sum_{i=1}^{n}\mathbf{A}^{i-1}\frac{d\mathbf{A}}{dt}\mathbf{A}^{n-i},$$

$$\frac{d\mathbf{A}^{-n}}{dt} = -\sum_{i=1}^{n}\mathbf{A}^{-i}\frac{d\mathbf{A}}{dt}\mathbf{A}^{-(n+1-i)}.$$

Therefore,

$$(\mathbf{A} + \epsilon\mathbf{X})^n = \mathbf{A}^n + \epsilon\sum_{i=1}^{n}\mathbf{A}^{i-1}\mathbf{X}\mathbf{A}^{n-i} + \mathcal{O}(\epsilon^2),$$

$$(\mathbf{A} + \epsilon\mathbf{X})^{-n} = \mathbf{A}^{-n} - \epsilon\sum_{i=1}^{n}\mathbf{A}^{-i}\mathbf{X}\mathbf{A}^{-(n+1-i)} + \mathcal{O}(\epsilon^2).$$

Generalized Inverse

Some of the properties of inverse matrices are shared by generalized inverses (for example, the Moore–Penrose inverse), which can be defined for any m-by-n matrix.

Applications

For most practical applications, it is *not* necessary to invert a matrix to solve a system of linear equations; however, for a unique solution, it *is* necessary that the matrix involved be invertible.

Decomposition techniques like LU decomposition are much faster than inversion, and various fast algorithms for special classes of linear systems have also been developed.

Regression/Least Squares

Although an explicit inverse is not necessary to estimate the vector of unknowns, it is the easiest way to estimate their accuracy, found in the diagonal of a matrix inverse (the posterior covariance matrix of the vector of unknowns). However, faster algorithms to compute only the diagonal entries of a matrix inverse are known in many cases.

Matrix Inverses in Real-time Simulations

Matrix inversion plays a significant role in computer graphics, particularly in 3D graphics rendering and 3D simulations. Examples include screen-to-world ray casting, world-to-subspace-to-world object transformations, and physical simulations.

Matrix Inverses in MIMO Wireless Communication

Matrix inversion also plays a significant role in the MIMO (Multiple-Input, Multiple-Output) technology in wireless communications. The MIMO system consists of N transmit and M receive antennas. Unique signals, occupying the same frequency band, are sent via N transmit antennas and are received via M receive antennas. The signal arriving at each receive antenna will be a linear combination of the N transmitted signals forming a NxM transmission matrix H. It is crucial for the matrix H to be invertible for the receiver to be able to figure out the transmitted information.

GAUSSIAN ELIMINATION METHOD

Gaussian elimination, also known as row reduction, is an algorithm in linear algebra for solving a system of linear equations. It is usually understood as a sequence of operations performed on the corresponding matrix of coefficients. This method can also be used to find the rank of a matrix, to calculate the determinant of a matrix, and to calculate the inverse of an invertible square matrix. The method is named after Carl Friedrich Gauss, although it was known to Chinese mathematicians as early as 179 AD.

To perform row reduction on a matrix, one uses a sequence of elementary row operations to modify the matrix until the lower left-hand corner of the matrix is filled with zeros, as much as possible. There are three types of elementary row operations:

- Swapping two rows,
- Multiplying a row by a nonzero number,
- Adding a multiple of one row to another row.

Using these operations, a matrix can always be transformed into an upper triangular matrix, and in fact one that is in row echelon form. Once all of the leading coefficients (the leftmost nonzero entry in each row) are 1, and every column containing a leading coefficient has zeros elsewhere, the matrix is said to be in reduced row echelon form. This final form is unique; in other words, it is independent of the sequence of row operations used. For example, in the following sequence of row operations (where multiple elementary operations might be done at each step), the third and fourth matrices are the ones in row echelon form, and the final matrix is the unique reduced row echelon form.

$$
\begin{bmatrix} 1 & 3 & 1 & 9 \\ 1 & 1 & -1 & 1 \\ 3 & 11 & 5 & 35 \end{bmatrix} \rightarrow
\begin{bmatrix} 1 & 3 & 1 & 9 \\ 0 & -2 & -2 & -8 \\ 0 & 2 & 2 & 8 \end{bmatrix} \rightarrow
\begin{bmatrix} 1 & 3 & 1 & 9 \\ 0 & -2 & -2 & -8 \\ 0 & 0 & 0 & 0 \end{bmatrix} \rightarrow
\begin{bmatrix} 1 & 0 & -2 & -3 \\ 0 & 1 & 1 & 4 \\ 0 & 0 & 0 & 0 \end{bmatrix}
$$

Using row operations to convert a matrix into reduced row echelon form is sometimes called Gauss–Jordan elimination. Some authors use the term Gaussian elimination to refer to the process until it has reached its upper triangular, or (unreduced) row echelon form. For computational reasons, when solving systems of linear equations, it is sometimes preferable to stop row operations before the matrix is completely reduced.

The process of row reduction makes use of elementary row operations, and can be divided into two parts. The first part (sometimes called forward elimination) reduces a given system to *row echelon form*, from which one can tell whether there are no solutions, a unique solution, or infinitely many solutions. The second part (sometimes called back substitution) continues to use row operations until the solution is found; in other words, it puts the matrix into *reduced* row echelon form.

Another point of view, which turns out to be very useful to analyze the algorithm, is that row reduction produces a matrix decomposition of the original matrix. The elementary row operations may be viewed as the multiplication on the left of the original matrix by elementary matrices. Alternatively, a sequence of elementary operations that reduces a single row may be viewed as multiplication by a Frobenius matrix. Then the first part of the algorithm computes an LU decomposition, while the second part writes the original matrix as the product of a uniquely determined invertible matrix and a uniquely determined reduced row echelon matrix.

Row Operations

There are three types of elementary row operations which may be performed on the rows of a matrix:

- Swap the positions of two rows.

- Multiply a row by a non-zero scalar.

- Add to one row a scalar multiple of another.

If the matrix is associated to a system of linear equations, then these operations do not change the solution set. Therefore, if one's goal is to solve a system of linear equations, then using these row operations could make the problem easier.

Echelon Form

For each row in a matrix, if the row does not consist of only zeros, then the leftmost nonzero entry is called the *leading coefficient* (or *pivot*) of that row. So if two leading coefficients are in the same column, then a row operation of type 3 could be used to make one of those coefficients zero. Then by using the row swapping operation, one can always order the rows so that for every non-zero row, the leading coefficient is to the right of the leading coefficient of the row above. If this is the case, then matrix is said to be in row echelon form. So the lower left part of the matrix contains only zeros, and all of the zero rows are below the non-zero rows. The word "echelon" is used here because one can roughly think of the rows being ranked by their size, with the largest being at the top and the smallest being at the bottom.

For example, the following matrix is in row echelon form, and its leading coefficients are shown in red:

$$\begin{bmatrix} 0 & 2 & 1 & -1 \\ 0 & 0 & 3 & 1 \\ 0 & 0 & 0 & 0 \end{bmatrix}.$$

It is in echelon form because the zero row is at the bottom, and the leading coefficient of the second row (in the third column), is to the right of the leading coefficient of the first row (in the second column).

A matrix is said to be in reduced row echelon form if furthermore all of the leading coefficients are equal to 1 (which can be achieved by using the elementary row operation of type 2), and in every column containing a leading coefficient, all of the other entries in that column are zero (which can be achieved by using elementary row operations of type 3).

Example of the Algorithm

Suppose the goal is to find and describe the set of solutions to the following system of linear equations:

$$2x + y - z = 8 \ (L_1)$$
$$-3x - y + 2z = -11 \ (L_2)$$
$$-2x + y + 2z = -3 \ (L_3)$$

The table below is the row reduction process applied simultaneously to the system of equations and its associated augmented matrix. In practice, one does not usually deal with the systems in terms of equations, but instead makes use of the augmented matrix, which is more suitable for computer manipulations. The row reduction procedure may be summarized as follows: eliminate x from all equations below L_1, and then eliminate y from all equations below L_2. This will put the system into triangular form. Then, using back-substitution, each unknown can be solved for.

System of equations	Row operations	Augmented matrix
$2x + y - z = 8$ $-3x - y + 2z = -11$ $-2x + y + 2z = -3$		$\begin{bmatrix} 2 & 1 & -1 & 8 \\ -3 & -1 & 2 & -11 \\ -2 & 1 & 2 & -3 \end{bmatrix}$
$2x + y - z = 8$ $\frac{1}{2}y + \frac{1}{2}z = 1$ $2y + z = 5$	$L_2 + \frac{3}{2}L_1 \to L_2$ $L_3 + L_1 \to L_3$	$\begin{bmatrix} 2 & 1 & -1 & 8 \\ 0 & \frac{1}{2} & \frac{1}{2} & 1 \\ 0 & 2 & 1 & 5 \end{bmatrix}$
$2x + y - z = 8$ $\frac{1}{2}y + \frac{1}{2}z = 1$ $-z = 1$	$L_3 + -4L_2 \to L_3$	$\begin{bmatrix} 2 & 1 & -1 & 8 \\ 0 & \frac{1}{2} & \frac{1}{2} & 1 \\ 0 & 0 & -1 & 1 \end{bmatrix}$
The matrix is now in echelon form (also called triangular form)		
$2x + y = 7$ $\frac{1}{2}y = \frac{3}{2}$ $-z = 1$	$L_2 + \frac{1}{2}L_3 \to L_2$ $L_1 - L_3 \to L_1$	$\begin{bmatrix} 2 & 1 & 0 & 7 \\ 0 & \frac{1}{2} & 0 & \frac{3}{2} \\ 0 & 0 & -1 & 1 \end{bmatrix}$
$2x + y = 7$ $y = 3$ $z = -1$	$2L_2 \to L_2$ $-L_3 \to L_3$	$\begin{bmatrix} 2 & 1 & 0 & 7 \\ 0 & 1 & 0 & 3 \\ 0 & 0 & 1 & -1 \end{bmatrix}$
$x = 2$ $y = 3$ $z = -1$	$L_1 - L_2 \to L_1$ $\frac{1}{2}L_1 \to L_1$	$\begin{bmatrix} 1 & 0 & 0 & 2 \\ 0 & 1 & 0 & 3 \\ 0 & 0 & 1 & -1 \end{bmatrix}$

The second column describes which row operations have just been performed. So for the first step, the x is eliminated from L_2 by adding $3/2L_1$ to L_2. Next, x is eliminated from L_3 by adding L_1 to L_3. These row operations are labelled in the table as:

$$L_2 + \tfrac{3}{2}L_1 \to L_2,$$
$$L_3 + L_1 \to L_3.$$

Once y is also eliminated from the third row, the result is a system of linear equations in triangular form, and so the first part of the algorithm is complete. From a computational point of view, it is faster to solve the variables in reverse order, a process known as back-substitution. One sees the solution is $z = -1$, $y = 3$, and $x = 2$. So there is a unique solution to the original system of equations.

Instead of stopping once the matrix is in echelon form, one could continue until the matrix is in *reduced* row echelon form, as it is done in the table. The process of row reducing until the matrix is reduced is sometimes referred to as Gauss–Jordan elimination, to distinguish it from stopping after reaching echelon form.

Applications

Historically, the first application of the row reduction method is for solving systems of linear equations. Here are some other important applications of the algorithm.

Computing Determinants

To explain how Gaussian elimination allows the computation of the determinant of a square matrix, we have to recall how the elementary row operations change the determinant:

- Swapping two rows multiplies the determinant by -1.

- Multiplying a row by a nonzero scalar multiplies the determinant by the same scalar.

- Adding to one row a scalar multiple of another does not change the determinant.

If Gaussian elimination applied to a square matrix A produces a row echelon matrix B, let d be the product of the scalars by which the determinant has been multiplied, using the above rules. Then the determinant of A is the quotient by d of the product of the elements of the diagonal of B:

$$\det(A) = \frac{\prod \mathrm{diag}(B)}{d}.$$

Computationally, for an $n \times n$ matrix, this method needs only $O(n^3)$ arithmetic operations, while solving by elementary methods requires $O(2^n)$ or $O(n!)$ operations. Even on the fastest computers, the elementary methods are impractical for n above 20.

Finding the Inverse of a Matrix

A variant of Gaussian elimination called Gauss–Jordan elimination can be used for finding the inverse of a matrix, if it exists. If A is an $n \times n$ square matrix, then one can use row reduction to compute its inverse matrix, if it exists. First, the $n \times n$ identity matrix is augmented to the right of A, forming an $n \times 2n$ block matrix $[A \mid I]$. Now through application of elementary row operations, find the reduced echelon form of this $n \times 2n$ matrix. The matrix A is invertible if and only if the left block can be reduced to the identity matrix I; in this case the right block of the final matrix is A^{-1}. If the algorithm is unable to reduce the left block to I, then A is not invertible. For example, consider the following matrix:

$$A = \begin{bmatrix} 2 & -1 & 0 \\ -1 & 2 & -1 \\ 0 & -1 & 2 \end{bmatrix}.$$

To find the inverse of this matrix, one takes the following matrix augmented by the identity and row-reduces it as a 3×6 matrix:

$$[A \mid I] = \begin{bmatrix} 2 & -1 & 0 & 1 & 0 & 0 \\ -1 & 2 & -1 & 0 & 1 & 0 \\ 0 & -1 & 2 & 0 & 0 & 1 \end{bmatrix}.$$

By performing row operations, one can check that the reduced row echelon form of this augmented matrix is:

$$[I \mid B] = \begin{bmatrix} 1 & 0 & 0 & \dfrac{3}{4} & \dfrac{1}{2} & \dfrac{1}{4} \\ 0 & 1 & 0 & \dfrac{1}{2} & 1 & \dfrac{1}{2} \\ 0 & 0 & 1 & \dfrac{1}{4} & \dfrac{1}{2} & \dfrac{3}{4} \end{bmatrix}.$$

One can think of each row operation as the left product by an elementary matrix. Denoting by B the product of these elementary matrices, we showed, on the left, that $BA = I$, and therefore, $B = A^{-1}$. On the right, we kept a record of $BI = B$, which we know is the inverse desired. This procedure for finding the inverse works for square matrices of any size.

Computing Ranks and Bases

The Gaussian elimination algorithm can be applied to any $m \times n$ matrix A. In this way, for example, some 6×9 matrices can be transformed to a matrix that has a row echelon form like

$$T = \begin{bmatrix} a & * & * & * & * & * & * & * & * \\ 0 & 0 & b & * & * & * & * & * & * \\ 0 & 0 & 0 & c & * & * & * & * & * \\ 0 & 0 & 0 & 0 & 0 & 0 & d & * & * \\ 0 & 0 & 0 & 0 & 0 & 0 & 0 & 0 & e \\ 0 & 0 & 0 & 0 & 0 & 0 & 0 & 0 & 0 \end{bmatrix},$$

where the stars are arbitrary entries, and a, b, c, d, e are nonzero entries. This echelon matrix T contains a wealth of information about A: the rank of A is 5, since there are 5 nonzero rows in T; the vector space spanned by the columns of A has a basis consisting of its columns 1, 3, 4, 7 and 9 (the columns with a, b, c, d, e in T), and the stars show how the other columns of A can be written as linear combinations of the basis columns. This is a consequence of the distributivity of the dot product in the expression of a linear map as a matrix.

All of this applies also to the reduced row echelon form, which is a particular row echelon format.

LU DECOMPOSITIONS

In numerical analysis and linear algebra, lower–upper (LU) decomposition or factorization factors a matrix as the product of a lower triangular matrix and an upper triangular matrix. The product sometimes includes a permutation matrix as well. LU decomposition can be viewed as the matrix form of Gaussian elimination. Computers usually solve square systems of linear equations using

LU decomposition, and it is also a key step when inverting a matrix or computing the determinant of a matrix. LU decomposition was introduced by mathematician Tadeusz Banachiewicz in 1938.

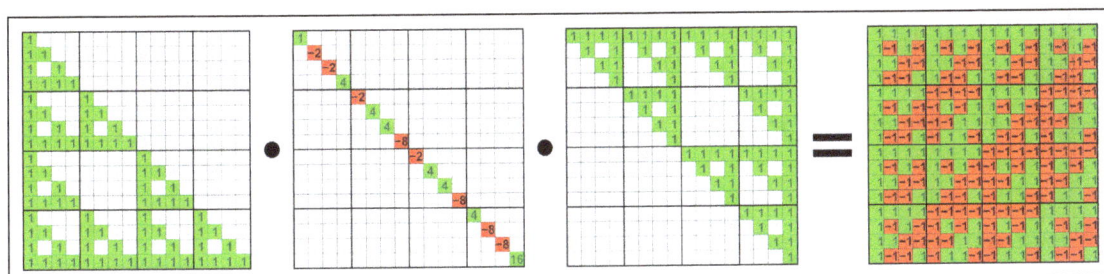

LDU decomposition of a Walsh matrix.

Let A be a square matrix. An LU factorization refers to the factorization of A, with proper row and/ or column orderings or permutations, into two factors – a lower triangular matrix L and an upper triangular matrix U:

$$A = LU.$$

In the lower triangular matrix all elements above the diagonal are zero, in the upper triangular matrix, all the elements below the diagonal are zero. For example, for a 3×3 matrix A, its LU decomposition looks like this:

$$\begin{bmatrix} a_{11} & a_{12} & a_{13} \\ a_{21} & a_{22} & a_{23} \\ a_{31} & a_{32} & a_{33} \end{bmatrix} = \begin{bmatrix} l_{11} & 0 & 0 \\ l_{21} & l_{22} & 0 \\ l_{31} & l_{32} & l_{33} \end{bmatrix} \begin{bmatrix} u_{11} & u_{12} & u_{13} \\ 0 & u_{22} & u_{23} \\ 0 & 0 & u_{33} \end{bmatrix}.$$

Without a proper ordering or permutations in the matrix, the factorization may fail to materialize. For example, it is easy to verify (by expanding the matrix multiplication) that $a_{11} = l_{11}u_{11}$. If $a_{11} = 0$, then at least one of l_{11} and u_{11} has to be zero, which implies that either L or U is singular. This is impossible if A is nonsingular (invertible). This is a procedural problem. It can be removed by simply reordering the rows of A so that the first element of the permuted matrix is nonzero. The same problem in subsequent factorization steps can be removed the same way.

LU Factorization with Partial Pivoting

It turns out that a proper permutation in rows (or columns) is sufficient for LU factorization. LU factorization with partial pivoting (LUP) refers often to LU factorization with row permutations only:

$$PA = LU,$$

where L and U are again lower and upper triangular matrices, and P is a permutation matrix, which, when left-multiplied to A, reorders the rows of A. It turns out that all square matrices can be factorized in this form, and the factorization is numerically stable in practice. This makes LUP decomposition a useful technique in practice.

LU Factorization with Full Pivoting

An LU factorization with full pivoting involves both row and column permutations:

$$PAQ = LU,$$

where L, U and P are defined as before, and Q is a permutation matrix that reorders the columns of A.

LDU Decomposition

An LDU decomposition is a decomposition of the form:

$$A = LDU,$$

where D is a diagonal matrix, and L and U are *unit* triangular matrices, meaning that all the entries on the diagonals of L and U are one.

Above we required that A be a square matrix, but these decompositions can all be generalized to rectangular matrices as well. In that case, L and D are square matrices both of which have the same number of rows as A, and U has exactly the same dimensions as A. *Upper triangular* should be interpreted as having only zero entries below the main diagonal, which starts at the upper left corner.

We factorize the following 2-by-2 matrix:

$$\begin{bmatrix} 4 & 3 \\ 6 & 3 \end{bmatrix} = \begin{bmatrix} l_{11} & 0 \\ l_{21} & l_{22} \end{bmatrix} \begin{bmatrix} u_{11} & u_{12} \\ 0 & u_{22} \end{bmatrix}.$$

One way to find the LU decomposition of this simple matrix would be to simply solve the linear equations by inspection. Expanding the matrix multiplication gives:

$$l_{11} \cdot u_{11} + 0 \cdot 0 = 4$$
$$l_{11} \cdot u_{12} + 0 \cdot u_{22} = 3$$
$$l_{21} \cdot u_{11} + l_{22} \cdot 0 = 6$$
$$l_{21} \cdot u_{12} + l_{22} \cdot u_{22} = 3.$$

This system of equations is underdetermined. In this case any two non-zero elements of L and U matrices are parameters of the solution and can be set arbitrarily to any non-zero value. Therefore, to find the unique LU decomposition, it is necessary to put some restriction on L and U matrices. For example, we can conveniently require the lower triangular matrix L to be a unit triangular matrix (i.e. set all the entries of its main diagonal to ones). Then the system of equations has the following solution:

$$l_{21} = 1.5$$
$$u_{11} = 4$$
$$u_{12} = 3$$
$$u_{22} = -1.5$$

Substituting these values into the LU decomposition above yields,

$$\begin{bmatrix} 4 & 3 \\ 6 & 3 \end{bmatrix} = \begin{bmatrix} 1 & 0 \\ 1.5 & 1 \end{bmatrix} \begin{bmatrix} 4 & 3 \\ 0 & -1.5 \end{bmatrix}.$$

Existence and Uniqueness

Square Matrices

Any square matrix A admits an *LUP* factorization. If A is invertible, then it admits an LU (or LDU) factorization if and only if all its leading principal minors are nonzero. If A is a singular matrix of rank k, then it admits an LU factorization if the first   leading principal minors are nonzero, although the converse is not true.

If a square, invertible matrix has an LDU factorization with all diagonal entries of L and U equal to 1, then the factorization is unique. In that case, the LU factorization is also unique if we require that the diagonal of L (or U) consists of ones.

Symmetric Positive Definite Matrices

If A is a symmetric (or Hermitian, if A is complex) positive definite matrix, we can arrange matters so that U is the conjugate transpose of L. That is, we can write A as:

$$A = LL^*.$$

This decomposition is called the Cholesky decomposition. The Cholesky decomposition always exists and is unique — provided the matrix is positive definite. Furthermore, computing the Cholesky decomposition is more efficient and numerically more stable than computing some other LU decompositions.

General Matrices

For a (not necessarily invertible) matrix over any field, the exact necessary and sufficient conditions under which it has an LU factorization are known. The conditions are expressed in terms of the ranks of certain submatrices. The Gaussian elimination algorithm for obtaining LU decomposition has also been extended to this most general case.

Algorithms

LU decomposition is basically a modified form of Gaussian elimination. We transform the matrix A into an upper triangular matrix U by eliminating the entries below the main diagonal. The Doolittle algorithm does the elimination column-by-column, starting from the left, by multiplying A to the left with atomic lower triangular matrices. It results in a *unit lower triangular* matrix and an upper triangular matrix. The Crout algorithm is slightly different and constructs a lower triangular matrix and a *unit upper triangular* matrix.

Computing an LU decomposition using either of these algorithms requires $2n^3/3$ floating-point operations, ignoring lower-order terms. Partial pivoting adds only a quadratic term; this is not the case for full pivoting.

Closed Formula

When an LDU factorization exists and is unique, there is a closed (explicit) formula for the elements of L, D, and U in terms of ratios of determinants of certain submatrices of the original matrix A. In particular, $D_1 = A_1$, and for $i = 2, \ldots, n$, D_i is the ratio of the i-th principal submatrix to the $(i-1)$-th principal submatrix. Computation of the determinants is computationally expensive, so this explicit formula is not used in practice.

Doolittle Algorithm

Given an $N \times N$ matrix,

$$A = (a_{i,j})_{1 \le i, j \le N},$$

we define,

$$A^{(0)} := A$$

We eliminate the matrix elements below the main diagonal in the n-th column of $A^{(n-1)}$ by adding to the i-th row of this matrix the n-th row multiplied by:

$$-l_{i,n} := -\frac{a_{i,n}^{(n-1)}}{a_{n,n}^{(n-1)}}$$

for $i = n+1, \ldots, N$. This can be done by multiplying $A^{(n-1)}$ to the left with the lower triangular matrix:

$$L_n = \begin{pmatrix} 1 & 0 & & \cdots & & 0 \\ 0 & \ddots & & \ddots & & \\ & & 1 & & & \\ \vdots & & -l_{n+1,n} & & & \vdots \\ & & \vdots & & \ddots & 0 \\ 0 & & -l_{N,n} & & & 1 \end{pmatrix}.$$

We set,

$$A^{(n)} := L_n A^{(n-1)}$$

After $N-1$ steps, we eliminated all the matrix elements below the main diagonal, so we obtain an upper triangular matrix $A^{(N-1)}$. We find the decomposition:

$$A = L_1^{-1} L_1 A^{(0)} = L_1^{-1} A^{(1)} = L_1^{-1} L_2^{-1} L_2 A^{(1)} = L_1^{-1} L_2^{-1} A^{(2)} = \cdots = L_1^{-1} \ldots L_{N-1}^{-1} A^{(N-1)}.$$

Denote the upper triangular matrix $A^{(N-1)}$ by U, and $L = L_1^{-1} \ldots L_{N-1}^{-1}$. Because the inverse of a lower triangular matrix L_n is again a lower triangular matrix, and the multiplication of two lower triangular matrices is again a lower triangular matrix, it follows that L is a lower triangular matrix. Moreover, it can be seen that:

$$L = \begin{pmatrix} 1 & 0 & \cdots & & & 0 \\ l_{2,1} & \ddots & \ddots & & & \\ & & 1 & & & \\ \vdots & & l_{n+1,n} & \ddots & & \vdots \\ & & \vdots & & 1 & 0 \\ l_{N,1} & & l_{N,n} & & l_{N,N-1} & 1 \end{pmatrix}.$$

We obtain $A = LU$.

It is clear that in order for this algorithm to work, one needs to have $a_{n,n}^{(n-1)} \neq 0$ at each step. If this assumption fails at some point, one needs to interchange n-th row with another row below it before continuing. This is why an LU decomposition in general looks like $P^{-1}A = LU$.

Crout and LUP Algorithms

The LUP decomposition algorithm by Cormen et al. generalizes Crout matrix decomposition. It can be described as follows:

1. If A has a nonzero entry in its first row, then take a permutation matrix P_1 such that AP_1 has a nonzero entry in its upper left corner. Otherwise, take for P_1 the identity matrix. Let $A_1 = AP_1$.

2. Let A_2 be the matrix that one gets from A_1 by deleting both the first row and the first column. Decompose $A_2 = L_2 U_2 P_2$ recursively. Make L from L_2 by first adding a zero row above and then adding the first column of A_1 at the left.

3. Make U_3 from U_2 by first adding a zero row above and a zero column at the left and then replacing the upper left entry (which is 0 at this point) by 1. Make P_3 from P_2 in a similar manner and define $A_3 = A_1 / P_3 = AP_1 / P_3$. Let P be the inverse of P_1 / P_3.

4. At this point, A_3 is the same as LU_3, except (possibly) at the first row. If the first row of A is zero, then $A_3 = LU_3$, since both have first row zero, and $A = LU_3 P$ follows, as desired. Otherwise, A_3 and LU_3 have the same nonzero entry in the upper left corner, and $A_3 = LU_3 U_1$ for some upper triangular square matrix U_1 with ones on the diagonal (U_1 clears entries of LU_3 and adds entries of A_3 by way of the upper left corner). Now $A = LU_3 U_1 P$ is a decomposition of the desired form.

Randomized Algorithm

It is possible to find a low rank approximation to an LU decomposition using a randomized algorithm. Given an input matrix A and a desired low rank k, the randomized LU returns permutation matrices P, Q and lower/upper trapezoidal matrices L, U of size $m \times k$ and $k \times n$ respectively, such that with high probability $\| PAQ - LU \|_2 \le C\sigma_{k+1}$, where C is a constant that depends on the parameters of the algorithm and   is the $(k+1)$ th singular value of the input matrix A.

Theoretical Complexity

If two matrices of order n can be multiplied in time $M(n)$, where $M(n) \ge n^a$ for some $a > 2$, then an LU decomposition can be computed in time $O(M(n))$. This means, for example, that an $O(n^{2.376})$ algorithm exists based on the Coppersmith–Winograd algorithm.

Sparse-matrix Decomposition

Special algorithms have been developed for factorizing large sparse matrices. These algorithms attempt to find sparse factors L and U. Ideally, the cost of computation is determined by the number of nonzero entries, rather than by the size of the matrix.

These algorithms use the freedom to exchange rows and columns to minimize fill-in (entries that change from an initial zero to a non-zero value during the execution of an algorithm).

General treatment of orderings that minimize fill-in can be addressed using graph theory.

Applications

Solving Linear Equations

Given a system of linear equations in matrix form:

$$Ax = b,$$

we want to solve the equation for x, given A and b. Suppose we have already obtained the LUP decomposition of A such that $PA = LU$, so $LUx = Pb$.

In this case the solution is done in two logical steps:

1.   First, we solve the equation $Ly = Pb$ for y.

2.   Second, we solve the equation $Ux = y$ for x.

In both cases we are dealing with triangular matrices (L and U), which can be solved directly by forward and backward substitution without using the Gaussian elimination process (however we do need this process or equivalent to compute the LU decomposition itself).

The above procedure can be repeatedly applied to solve the equation multiple times for different b. In this case it is faster (and more convenient) to do an LU decomposition of the matrix A once and then solve the triangular matrices for the different b, rather than using Gaussian elimination each time. The matrices L and U could be thought to have "encoded" the Gaussian elimination process.

The cost of solving a system of linear equations is approximately $\frac{2}{3}n^3$ floating-point operations if the matrix A has size n. This makes it twice as fast as algorithms based on QR decomposition, which costs about $\frac{4}{3}n^3$ floating-point operations when Householder reflections are used. For this reason, LU decomposition is usually preferred.

Inverting a Matrix

When solving systems of equations, b is usually treated as a vector with a length equal to the height of matrix A. In matrix inversion however, instead of vector b, we have matrix B, where B is an n-by-p matrix, so that we are trying to find a matrix X (also a n-by-p matrix):

$AX = LUX = B.$

We can use the same algorithm presented earlier to solve for each column of matrix X. Now suppose that B is the identity matrix of size n. It would follow that the result X must be the inverse of A.

Computing the Determinant

Given the LUP decomposition $A = P^{-1}LU$ of a square matrix A, the determinant of A can be computed straightforwardly as:

$$\det(A) = \det(P^{-1})\det(L)\det(U) = (-1)^S \left(\prod_{i=1}^{n} l_{ii} \right)\left(\prod_{i=1}^{n} u_{ii} \right).$$

The second equation follows from the fact that the determinant of a triangular matrix is simply the product of its diagonal entries, and that the determinant of a permutation matrix is equal to $(-1)^S$ where S is the number of row exchanges in the decomposition.

In the case of LU decomposition with full pivoting, $\det(A)$ also equals the right-hand side of the above equation, if we let S be the total number of row and column exchanges.

The same method readily applies to LU decomposition by setting P equal to the identity matrix.

EIGEN VALUES AND EIGEN VECTORS OF A MATRIX

Let A be an $n \times n$ matrix. The number λ is an eigenvalue of A if there exists a non-zero vector v such that:

$Av = \lambda v.$

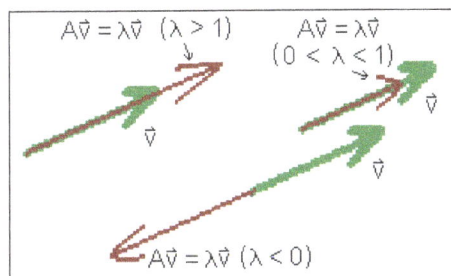

In this case, vector v is called an eigenvector of A corresponding to λ.

Computing Eigenvalues and Eigenvectors

We can rewrite the condition $Av = \lambda v$ as:

$$(A - \lambda I)v = 0$$

where I is the $n \times n$ identity matrix. Now, in order for a *non-zero* vector **v** to satisfy this equation, $A - \lambda I$ must *not* be invertible.

Otherwise, if $A - \lambda I$ has an inverse:

$$(A - \lambda I)^{-1}(A - \lambda I)v = (A - \lambda I)^{-1}0$$
$$v = 0.$$

But we are looking for a non-zero vector v.

That is, the determinant of $A - \lambda I$ must equal 0. We call $p(\lambda) = \det(A - \lambda I)$ the characteristic polynomial of A. The eigenvalues of A are simply the roots of the characteristic polynomial of A.

Example:

Let $A = \begin{bmatrix} 2 & -4 \\ -1 & -1 \end{bmatrix}$

Then,

$$\begin{aligned} p &= \det \begin{bmatrix} 2 & -4 \\ -1 & -1-\lambda \end{bmatrix} \\ &= (2-\lambda)(-1-\lambda) - (-4)(-1) \\ &= \lambda^2 - \lambda - 6 \\ &= (\lambda - 3)(\lambda + 2). \end{aligned}$$

Thus, $\lambda_1 = 3$ and $\lambda_2 = -2$ are the eigenvalues of A.

To find eigenvectors $v = \begin{bmatrix} v_1 \\ v_2 \\ \vdots \\ v_n \end{bmatrix}$ corresponding to an eigenvalue λ, we simply solve the system of linear equations given by:

$$(A - \lambda I)v = 0$$

Example:

$A = \begin{bmatrix} 2 & -4 \\ -1 & -1 \end{bmatrix}$ of the previous examples has eigenvlues $\lambda_1 = 3$ and $\lambda_2 = -2$.

Let's find the eigenvectors corresponding to $\lambda_1 = 3$. Let $v = \begin{bmatrix} v_1 \\ v_2 \end{bmatrix}$. Then $(A - 3I)v = 0$ given us:

$$\begin{bmatrix} 2-3 & -4 \\ -1 & -1-3 \end{bmatrix} \begin{bmatrix} v_1 \\ v_2 \end{bmatrix} = \begin{bmatrix} 0 \\ 0 \end{bmatrix},$$

from which we obtain the duplicate equations:

$$-v_1 - 4v_2 = 0$$
$$-v_1 - 4v_2 = 0.$$

If we let $v_2 = t$, then $v_1 = -4t$. All eigenvectors corresponding to $\lambda_1 = 3$ are multiples of $\begin{bmatrix} -4 \\ 1 \end{bmatrix}$ and thus

the eigenspace corresponding to $\lambda_1 = 3$ is given by the span of $\begin{bmatrix} -4 \\ 1 \end{bmatrix}$ That is, $\left\{ \begin{bmatrix} -4 \\ 1 \end{bmatrix} \right\}$ is a basis of

the eigenspace corresponding to $\lambda_1 = 3$.

Repeating this process with $\lambda_2 = -2$, we find $\begin{array}{c} 4v_1 - 4V_2 = 0 \\ -v_1 + v_2 = 0 \end{array}$.

If we let $v_2 = t$ then $v_1 = t$ as well. Thus, an eigenvector corresponding to $\lambda_2 = -2$ is $\begin{bmatrix} 1 \\ 1 \end{bmatrix}$ and the ei-

genspace corresponding to $\lambda_2 = -2$ is given by the span of $\begin{bmatrix} 1 \\ 1 \end{bmatrix}. \left\{ \begin{bmatrix} 1 \\ 1 \end{bmatrix} \right\}$ is a basis for the eigenspace

corresponding to $\lambda_2 = -2$.

In the following example, we see a two-dimensional eigenspace.

Example:

Let $A = \begin{bmatrix} 5 & 8 & 16 \\ 4 & 1 & 8 \\ -4 & -4 & -11 \end{bmatrix}$. Then $p(\lambda) = \det \begin{bmatrix} 5-\lambda & 8 & 16 \\ 4 & 1-\lambda & 8 \\ -4 & -4 & -11-\lambda \end{bmatrix} = (\lambda-1)(\lambda+3)^2$

Thus, $\lambda_1 = 1$ and $\lambda_2 = -3$ are the eigenvalues of A. Eigenvectors $v = \begin{bmatrix} v_1 \\ v_2 \\ v_3 \end{bmatrix}$ corresponding to $\lambda_1 = 1$
must satisfy:

$$4v_1 + 8v_2 + 16v_3 = 0$$
$$4v_1 + 4v_1 + 8v_3 = 0$$
$$-4v_1 - 4v_2 - 8v_3 = 0.$$

Letting $v_3 = t$, we find from the second equation that $v_1 = -2t$, and then $v_2 = -t$. All eigenvectors

corresponding to $\lambda_1 = 1$ are multiples of $\begin{bmatrix} -2 \\ -1 \\ 1 \end{bmatrix}$ and so the eigenspace corresponding to $\lambda_1 = 1$ is given by the span of $\left\{ \begin{bmatrix} -2 \\ -1 \\ 1 \end{bmatrix}, \begin{bmatrix} -2 \\ -1 \\ 1 \end{bmatrix} \right\}$ is a basis for the eigenspace corresponding to $\lambda_1 = 1$.

Eigenvectors corresponding to $\lambda_2 = -3$ must satisfy:

$$4v_1 + 8v_2 + 16v_3 = 0$$
$$4v_1 + 4v_1 + 8v_3 = 0$$
$$-4v_1 - 4v_2 - 8v_3 = 0.$$

The equations here are just multiples of each other. If we let $v_3=t$ and $v_2=s$, then $v_1=-s-2t$. Eigenvectors corresponding to $\lambda_2 = -3$ have the form:

$$\begin{bmatrix} -1 \\ 1 \\ 0 \end{bmatrix} s + \begin{bmatrix} -2 \\ 0 \\ 1 \end{bmatrix} t.$$

Thus, the eigenspace corresponding to $\lambda_2 = -3$ is two-dimensional and is spanned by:

$$\begin{bmatrix} -1 \\ 1 \\ 0 \end{bmatrix} \text{ and } \begin{bmatrix} -2 \\ 0 \\ 1 \end{bmatrix}. \left\{ \begin{bmatrix} -1 \\ 1 \\ 0 \end{bmatrix} \begin{bmatrix} -2 \\ 0 \\ 1 \end{bmatrix} \right\}$$ is a basis for the eigenspace corresponding to $\lambda_2 = -3$.

CAYLEY-HAMILTON THEOREM

Arthur Cayley, F.R.S. is widely regarded as Britain's leading pure mathematician of the 19th century. Cayley in 1848 went to Dublin to attend lectures on quaternions by Hamilton, their discoverer. Later Cayley impressed him by being the second to publish work on them. Cayley proved the theorem for matrices of dimension 3 and less, publishing proof for the two-dimensional case. As for n × n matrices, Cayley stated "..., I have not thought it necessary to undertake the labor of a formal proof of the theorem in the general case of a matrix of any degree".

William Rowan Hamilton, Irish physicist, astronomer, and mathematician, first foreign member of the American National Academy of Sciences. While maintaining opposing position about how geometry should be studied, Hamilton always remained on the best terms with Cayley.

Hamilton proved that for a linear function of quaternions there exists a certain equation, depending on the linear function, that is satisfied by the linear function itself.

In linear algebra, the Cayley–Hamilton theorem (named after the mathematicians Arthur Cayley and William Rowan Hamilton) states that every square matrix over a commutative ring (such as the real or complex field) satisfies its own characteristic equation.

If A is a given $n \times n$ matrix and I_n is the $n \times n$ identity matrix, then the characteristic polynomial of A is defined as:

$$p(\lambda) = \det(\lambda I_n - A),$$

where det is the determinant operation and λ is a scalar element of the base ring. Since the entries of the matrix $(\lambda I_n - A)$ are (linear or constant) polynomials in λ, the determinant is also an n-th order monic polynomial in λ. The Cayley–Hamilton theorem states that if one defines an analogous matrix equation, $p(A)$, consisting of the replacement of the scalar eigenvalues λ with the matrix A, then this polynomial in the matrix A results in the zero matrix:

$$p(A) = 0.$$

The powers of A, obtained by substitution from powers of λ, are defined by repeated matrix multiplication; the constant term of $p(\lambda)$ gives a multiple of the power A^0, which is defined as the identity matrix. The theorem allows A^n to be expressed as a linear combination of the lower matrix powers of A. When the ring is a field, the Cayley–Hamilton theorem is equivalent to the statement that the minimal polynomial of a square matrix divides its characteristic polynomial.

The theorem was first proved in 1853 in terms of inverses of linear functions of quaternions, a *non-commutative* ring, by Hamilton. This corresponds to the special case of certain 4×4 real or 2×2 complex matrices. The theorem holds for general quaternionic matrices. Cayley in 1858 stated it for 3×3 and smaller matrices, but only published a proof for the 2×2 case. The general case was first proved by Frobenius in 1878.

Examples:

1×1 Matrices

For a 1×1 matrix $A = (a_{1,1})$, the characteristic polynomial is given by $p(\lambda) = \lambda - a$, and so $p(A) = (a) - a_{1,1} = 0$ is obvious.

2×2 Matrices

As a concrete example, let:

$$A = \begin{pmatrix} 1 & 2 \\ 3 & 4 \end{pmatrix}.$$

Its characteristic polynomial is given by:

$$p(\lambda) = \det(\lambda I_2 - A) = \det \begin{pmatrix} \lambda - 1 & -2 \\ -3 & \lambda - 4 \end{pmatrix} = (\lambda - 1)(\lambda - 4) - (-2)(-3) = \lambda^2 - 5\lambda - 2.$$

The Cayley–Hamilton theorem claims that, if we define,

$$p(X) = X^2 - 5X - 2I_2,$$

then,

$$p(A) = A^2 - 5A - 2I_2 = \begin{pmatrix} 0 & 0 \\ 0 & 0 \end{pmatrix}.$$

We can verify by computation that indeed,

$$A^2 - 5A - 2I_2 = \begin{pmatrix} 7 & 10 \\ 15 & 22 \end{pmatrix} - \begin{pmatrix} 5 & 10 \\ 15 & 20 \end{pmatrix} - \begin{pmatrix} 2 & 0 \\ 0 & 2 \end{pmatrix} = \begin{pmatrix} 0 & 0 \\ 0 & 0 \end{pmatrix}.$$

For a generic 2×2 matrix,

$$A = \begin{pmatrix} a & b \\ c & d \end{pmatrix},$$

the characteristic polynomial is given by $p(\lambda) = \lambda^2 - (a + d)\lambda + (ad - bc)$, so the Cayley–Hamilton theorem states that:

$$p(A) = A^2 - (a+d)A + (ad-bc)I_2 = \begin{pmatrix} 0 & 0 \\ 0 & 0 \end{pmatrix};$$

which is indeed always the case, evident by working out the entries of A^2.

Applications

Determinant and Inverse Matrix

For a general $n \times n$ invertible matrix A, i.e., one with nonzero determinant, A^{-1} can thus be written as an $(n - 1)$-th order polynomial expression in A: As indicated, the Cayley–Hamilton theorem amounts to the identity:

$$p(A) = A^n + c_{n-1}A^{n-1} + \cdots + c_1 A + (-1)^n \det(A)I_n = O.$$

The coefficients c_i are given by the elementary symmetric polynomials of the eigenvalues of A. Using Newton identities, the elementary symmetric polynomials can in turn be expressed in terms of power sum symmetric polynomials of the eigenvalues:

$$s_k = \sum_{i=1}^{n} \lambda_i^k = \mathrm{tr}(A^k),$$

where tr (A^k) is the trace of the matrix A^k. Thus, we can express c_i in terms of the trace of powers of A.

In general, the formula for the coefficients c_i is given in terms of complete exponential Bell polynomials as:

$$c_{n-k} = \frac{(-1)^k}{k!} B_k(s_1, -1!s_2, 2!s_3, \ldots, (-1)^{k-1}(k-1)!s_k).$$

In particular, the determinant of A corresponds to c_0. Thus, the determinant can be written as a trace identity:

$$\det(A) = \frac{1}{n!} B_n(s_1, -1!s_2, 2!s_3, \ldots, (-1)^{n-1}(n-1)!s_n).$$

Likewise, the characteristic polynomial can be written as:

$$-(-1)^n \det(A)I_n = A(A^{n-1} + c_{n-1}A^{n-2} + \cdots + c_1 I_n),$$

and, by multiplying both sides by A^{-1} (note $-(-1)^n = (-1)^{n-1}$), one is led to an expression for the inverse of A as a trace identity,

$$A^{-1} = \frac{(-1)}{\det}(A^{n-1} + c_{n-1}A^{n-2} + \; + c_1 I_n),$$

$$= \frac{1}{\det A} \sum (-1)^{n+k-1} \frac{\overset{n-k-}{}}{k!} B_k(s_1, -1!s_2, 2!s_3, \ldots, (-1)^{k-1}(k-1)!s_k).$$

For instance, the first few Bell polynomials are $B_0 = 1$, $B_1(x_1) = x_1$, $B_2(x_1, x_2) = x_1^2$

$1 + x_2$, and $B_3(x_1, x_2, x_3) = x_1^3$

$1 + 3\,x_1 x_2 + x_3$.

Using these to specify the coefficients c_i of the characteristic polynomial of a 2×2 matrix yields:

$$c_2 = B_0 = 1,$$

$$c_1 = \frac{-1}{1!} B_1(s_1) = -s_1 = -\operatorname{tr}(A),$$

$$c_0 = \frac{1}{2!} B_2(s_1, -1!s_2) = \frac{1}{2}(s_1^2 - s_2) = \frac{1}{2}((\operatorname{tr}(A))^2 - \operatorname{tr}(A^2))$$

The coefficient c_0 gives the determinant of the 2×2 matrix, c_1 minus its trace, while its inverse is given by:

$$A^{-1} = \frac{-1}{\det A}(A + c_1 I_2) = \frac{-2(A - \operatorname{tr}(A)I_2)}{(\operatorname{tr}(A))^2 - \operatorname{tr}(A^2)}.$$

It is apparent from the general formula for c_{n-k}, expressed in terms of Bell polynomials, that the expressions:

$$-\operatorname{tr}(A) \quad \text{and} \quad \tfrac{1}{2}(\operatorname{tr}(A)^2 - \operatorname{tr}(A^2))$$

always give the coefficients c_{n-1} of λ^{n-1} and c_{n-2} of λ^{n-2} in the characteristic polynomial of any $n\times n$ matrix, respectively. So, for a 3×3 matrix A, the statement of the Cayley–Hamilton theorem can also be written as:

$$A^3 - (\operatorname{tr} A)A^2 + \frac{1}{2}\left((\operatorname{tr} A)^2 - \operatorname{tr}(A^2)\right)A - \det(A)I_3 = O,$$

where the right-hand side designates a 3×3 matrix with all entries reduced to zero. Likewise, this determinant in the $n = 3$ case, is now:

$$\det(A) = \frac{1}{3!} B_3(s_1, -1!s_2, 2!s_3)$$

$$= \frac{1}{6}(s_1^3 + 3s_1(-s_2) + 2s_3) = \frac{1}{6}\left((\operatorname{tr} A)^3 - 3\operatorname{tr}(A^2)(\operatorname{tr} A) + 2\operatorname{tr}(A^3)\right).$$

This expression gives the negative of coefficient c_{n-3} of λ^{n-3} in the general case.

Similarly, one can write for a 4×4 matrix A,

$$A^4 - (\operatorname{tr} A)A^3 + \frac{1}{2}\left((\operatorname{tr} A)^2 - \operatorname{tr}(A^2)\right)A^2 - \frac{1}{6}\left((\operatorname{tr} A)^3 - 3\operatorname{tr}(A^2)(\operatorname{tr} A) + 2\operatorname{tr}(A^3)\right)A + \det(A)I_4 = O,$$

where, now, the determinant is c_{n-4},

$$\frac{1}{24}\left((\operatorname{tr} A)^4 - 6\operatorname{tr}(A^2)(\operatorname{tr} A)^2 + 3(\operatorname{tr}(A^2))^2 + 8\operatorname{tr}(A^3)\operatorname{tr}(A) - 6\operatorname{tr}(A^4)\right),$$

and so on for larger matrices. The increasingly complex expressions for the coefficients c_k is deducible from Newton's identities or the Faddeev–LeVerrier algorithm.

Another method for obtaining these coefficients c_k for a general $n \times n$ matrix, provided no root be zero, relies on the following alternative expression for the determinant,

$$p(\lambda) = \det(\lambda I_n - A) = \lambda^n \exp(\operatorname{tr}(\log(I_n - A/\lambda))).$$

Hence, by virtue of the Mercator series,

$$p(\lambda) = \lambda^n \exp\left(-\operatorname{tr} \sum_{m=1}^{\infty} \frac{\left(\frac{A}{\lambda}\right)^m}{m}\right),$$

where the exponential *only* needs be expanded to order λ^{-n}, since $p(\lambda)$ is of order n, the net negative powers of λ automatically vanishing by the C–H theorem. (Again, this requires a ring containing the rational numbers.) The coefficients of λ can be directly written in terms of complete Bell polynomials by comparing this expression with the generating function of the Bell polynomial.

Differentiation of this expression with respect to λ allows determination of the generic coefficients of the characteristic polynomial for general n, as determinants of $m \times m$ matrices,

$$c_{n-m} = \frac{(-1)^m}{m!} \begin{vmatrix} \operatorname{tr} A & m-1 & 0 & \cdots & \\ \operatorname{tr} A^2 & \operatorname{tr} A & m-2 & \cdots & \\ \vdots & \vdots & & & \vdots \\ \operatorname{tr} A^{m-1} & \operatorname{tr} A^{m-2} & \cdots & \cdots & 1 \\ \operatorname{tr} A^m & \operatorname{tr} A^{m-1} & \cdots & \cdots & \operatorname{tr} A \end{vmatrix}.$$

n-th Power of Matrix

The Cayley–Hamilton theorem always provides a relationship between the powers of A (though not always the simplest one), which allows one to simplify expressions involving such powers, and evaluate them without having to compute the power A^n or any higher powers of A.

As an example, for $A = \begin{pmatrix} 1 & 2 \\ 3 & 4 \end{pmatrix}$ the theorem gives:

$$A^2 = 5A + 2I_2.$$

Then, to calculate A^4, observe:

$$A^3 = (5A + 2I_2)A = 5A^2 + 2A = 5(5A + 2I_2) + 2A = 27A + 10I_2.$$

$$A^4 = A^3 A = (27A + 10I_2)A = 27A^2 + 10A = 27(5A + 2I_2) + 10A = 145A + 54I_2.$$

Likewise,

$$A^{-1} = \frac{A - 5I_2}{2}.$$

Notice that we have been able to write the matrix power as the sum of two terms. In fact, matrix power of any order k can be written as a matrix polynomial of degree at most $n - 1$, where n is the size of a square matrix. This is an instance where Cayley–Hamilton theorem can be used to express a matrix function, which we will discuss below systematically.

Matrix Functions

Given an analytic function,

$$f(x) = \sum_{k=0}^{\infty} a_k x^k$$

and the characteristic polynomial $p(x)$ of degree n of an $n \times n$ matrix A, the function can be expressed using long division as:

$$f(x) = q(x)p(x) + r(x),$$

where $q(x)$ is some quotient polynomial and $r(x)$ is a remainder polynomial such that $0 \le \deg r(x) < n$. By the Cayley–Hamilton theorem, replacing x by the matrix A gives $p(A) = 0$, so one has:

$$f(A) = r(A).$$

Thus, the analytic function of matrix A can be expressed as a matrix polynomial of degree less than n.

Let the remainder polynomial be:

$$r(x) = c_0 + c_1 x + \cdots + c_{n-1}x^{n-1}.$$

Since $p(\lambda) = 0$, evaluating the function $f(x)$ at the n eigenvalues of A, yields:

$$f(\lambda_i) = r(\lambda_i) = c_0 + c_1\lambda_i + \cdots + c_{n-1}\lambda_i^{n-1}, \qquad \text{for} \qquad i = 1, 2, \ldots, n.$$

This amounts to a system of n linear equations, which can be solved to determine the coefficients c_i. Thus, one has:

$$f(A) = \sum_{k=0}^{n-1} c_k A^k.$$

When the eigenvalues are repeated, that is $\lambda_i = \lambda_j$ for some $i \neq j$, two or more equations are identical; and hence the linear equations cannot be solved uniquely. For such cases, for an eigenvalue λ with multiplicity m, the first $m - 1$ derivative of $p(x)$ vanishes at the eigenvalues. Thus, there are the extra $m - 1$ linearly independent solutions:

$$\left.\frac{\mathrm{d}^k f(x)}{\mathrm{d}x^k}\right|_{x=\lambda} = \left.\frac{\mathrm{d}^k r(x)}{\mathrm{d}x^k}\right|_{x=\lambda} \qquad \text{for} \qquad k = 1, 2, \ldots, m-1,$$

which, when combined with others, yield the required n equations to solve for c_i.

Finding a polynomial that passes through the points $(\lambda_i, f(\lambda_i))$ is essentially an interpolation problem, and can be solved using Lagrange or Newton interpolation techniques, leading to Sylvester's formula.

For example, suppose the task is to find the polynomial representation of:

$$f(A) = e^{At} \qquad \text{where} \qquad A = \begin{pmatrix} 1 & 2 \\ 0 & 3 \end{pmatrix}.$$

The characteristic polynomial is $p(x) = (x - 1)(x - 3) = x^2 - 4x + 3$, and the eigenvalues are $\lambda = 1$, 3. Let $r(x) = c_0 + c_1 x$. Evaluating $f(\lambda) = r(\lambda)$ at the eigenvalues, one obtains two linear equations $e^t = c_0 + c_1$ and $e^{3t} = c_0 + 3c_1$. Solving the equations yields $c_0 = (3e^t - e^{3t})/2$ and $c_1 = (e^{3t} - e^t)/2$. Thus, it follows that:

$$e^{At} = c_0 I_2 + c_1 A = \begin{pmatrix} c_0 + c_1 & 2c_1 \\ 0 & c_0 + 3c_1 \end{pmatrix} = \begin{pmatrix} e^t & e^{3t} - e^t \\ 0 & e^{3t} \end{pmatrix}.$$

If, instead, the function were $f(A) = \sin At$, then the coefficients would have been $c_0 = (3\sin t - \sin 3t)/2$ and $c_1 = (\sin 3t - \sin t)/2$; hence,

$$\sin(At) = c_0 I_2 + c_1 A = \begin{pmatrix} \sin t & \sin 3t - \sin t \\ 0 & \sin 3t \end{pmatrix}.$$

As a further example, when considering:

$$f(A) = e^{At} \qquad \text{where} \qquad A = \begin{pmatrix} 0 & 1 \\ -1 & 0 \end{pmatrix},$$

then the characteristic polynomial is $p(x) = x^2 + 1$, and the eigenvalues are $\lambda = \pm i$. As before, evaluating the function at the eigenvalues gives us the linear equations $e^{it} = c_0 + i\,c_1$ and $e^{-it} = c_0 - ic_1$; the solution of which gives, $c_0 = (e^{it} + e^{-it})/2 = \cos t$ and $c_1 = (e^{it} - e^{-it})/2i = \sin t$. Thus, for this case,

$$e^{At} = (\cos t)I_2 + (\sin t)A = \begin{pmatrix} \cos t & \sin t \\ -\sin t & \cos t \end{pmatrix},$$

which is a rotation matrix.

Standard examples of such usage is the exponential map from the Lie algebra of a matrix Lie group into the group. It is given by a matrix exponential,

$$\exp : \mathfrak{g} \to G; \qquad tX \mapsto e^{tX} = \sum_{n=0}^{\infty} \frac{t^n X^n}{n!} = I + tX + \frac{t^2 X^2}{2} + \cdots, t \in \mathbb{R}, X \in \mathfrak{g}.$$

Such expressions have long been known for SU(2),

$$e^{i(\theta/2)(\hat{n}\cdot\sigma)} = I_2 \cos\theta/2 + i(\hat{n}\cdot\sigma)\sin\theta/2,$$

where the σ are the Pauli matrices and for SO(3),

$$e^{i\theta(\hat{n}\cdot\mathbf{J})} = I_3 + i(\hat{n}\cdot\mathbf{J})\sin\theta + (\hat{n}\cdot\mathbf{J})^2(\cos\theta - 1),$$

which is Rodrigues' rotation formula.

More recently, expressions have appeared for other groups, like the Lorentz group SO(3, 1), O(4, 2) and SU(2, 2), as well as GL(n, **R**). The group O(4, 2) is the conformal group of spacetime, SU(2, 2) its simply connected cover (to be precise, the simply connected cover of the connected component SO$^+$(4, 2) of O(4, 2)). The expressions obtained apply to the standard representation of these groups. They require knowledge of (some of) the eigenvalues of the matrix to exponentiate. For SU(2) (and hence for SO(3)), closed expressions have recently been obtained for all irreducible representations, i.e. of any spin.

Algebraic Number Theory

Ferdinand Georg Frobenius, German mathematician. His main interests were elliptic functions differential equations, and later group theory. In 1878 he gave the first full proof of the Cayley–Hamilton theorem.

The Cayley–Hamilton theorem is an effective tool for computing the minimal polynomial of algebraic integers. For example, given a finite extension $\mathbb{Q}[\alpha_1,\dots,\alpha_k]$ of $\mathbb{Q}$ and an algebraic integer $\alpha \in \mathbb{Q}[\alpha_1,\dots,\alpha_k]$ which is a non-zero linear combination of the $\alpha_1^{n_1}\cdots\alpha_k^{n_k}$ we can compute the minimal polynomial of α by finding a matrix representing the $\mathbb{Q}$-linear transformation

$$\cdot\alpha : \mathbb{Q}[\alpha_1,\dots,\alpha_k] \to \mathbb{Q}[\alpha_1,\dots,\alpha_k]$$

If we call this transformation matrix A, then we can find the minimal polynomial by applying the Cayley–Hamilton theorem to A.

Proving the Theorem in General

The Cayley–Hamilton theorem is an immediate consequence of the existence of the Jordan normal form for matrices over algebraically closed fields. In this topic direct proofs are presented.

As the examples above show, obtaining the statement of the Cayley–Hamilton theorem for an $n \times n$ matrix:

$$A = (a_{ij})_{i,j=1}^n$$

requires two steps: first the coefficients c_i of the characteristic polynomial are determined by development as a polynomial in t of the determinant:

$$p(t) = \det(tI_n - A) = \begin{vmatrix} t-a_{1,1} & -a_{1,2} & \cdots & -a_{1,n} \\ -a_{2,1} & t-a_{2,2} & \cdots & -a_{2,n} \\ \vdots & \vdots & \ddots & \vdots \\ -a_{n,1} & -a_{n,2} & \cdots & t-a_{n,n} \end{vmatrix}$$

$$= t^n + c_{n-1}t^{n-1} + \cdots + c_1 t + c_0,$$

and then these coefficients are used in a linear combination of powers of A that is equated to the $n \times n$ null matrix:

$$A^n + c_{n-1}A^{n-1} + \cdots + c_1 A + c_0 I_n = \begin{pmatrix} 0 & \cdots & 0 \\ \vdots & \ddots & \vdots \\ 0 & \cdots & 0 \end{pmatrix}.$$

The left hand side can be worked out to an $n \times n$ matrix whose entries are (enormous) polynomial expressions in the set of entries $a_{i,j}$ of A, so the Cayley–Hamilton theorem states that each of these n^2 expressions are equal to 0. For any fixed value of n these identities can be obtained by tedious but completely straightforward algebraic manipulations. None of these computations can show however why the Cayley–Hamilton theorem should be valid for matrices of all possible sizes n, so a uniform proof for all n is needed.

Preliminaries

If a vector v of size n happens to be an eigenvector of A with eigenvalue λ, in other words if $A \cdot v = \lambda v$, then:

$$p(A) \cdot v = A^n \cdot v + c_{n-1}A^{n-1} \cdot v + \cdots + c_1 A \cdot v + c_0 I_n \cdot v$$

$$= \lambda^n v + c_{n-1}\lambda^{n-1}v + \cdots + c_1 \lambda v + c_0 v = p(\lambda)v,$$

which is the null vector since $p(\lambda) = 0$ (the eigenvalues of A are precisely the roots of $p(t)$). This holds for all possible eigenvalues λ, so the two matrices equated by the theorem certainly give the same (null) result when applied to any eigenvector. Now if A admits a basis of eigenvectors, in other words if A is diagonalizable, then the Cayley–Hamilton theorem must hold for A, since two matrices that give the same values when applied to each element of a basis must be equal.

$$A = XDX^{-1}, \quad D = \mathrm{diag}(\lambda_i), \quad i = 1, 2, \dots, n$$

$$p_A(\lambda) = |\lambda I - A| = \text{product of eigenvalues of } \lambda I - A = \prod_{i=1}^{n}(\lambda - \lambda_i) \equiv \sum_{k=0}^{n} c_k \lambda^k$$

$$p_A(A) = \sum c_k A^k = X p_A(D) X^{-1} = XCX^{-1}$$

$$C_{ii} = \sum_{k=0}^{n} c_k \lambda_i^k = \prod_{j=1}^{n}(\lambda_i - \lambda_j) = 0, \qquad C_{i,j\neq i} = 0$$

$$\therefore p_A(A) = XCX^{-1} = O.$$

Consider now the function $e : M_n \to M_n$ which maps $n \times n$ matrices to $n \times n$ matrices given by the formula $e(A) = p_A(A)$, i.e. which takes a matrix A and plugs it into its own characteristic polynomial. Not all matrices are diagonalizable, but for matrices with complex coefficients many of them are: the set of D diagonalizable complex square matrices of a given size is dense in the set of all such square matrices (for a matrix to be diagonalizable it suffices for instance that its characteristic polynomial not have any multiple roots). Now viewed as a function $e : \mathbb{C}^{n^2} \to \mathbb{C}^{n^2}$ (since matrices have n^2 entries) we see that this function is continuous. This is true because the entries of the image of a matrix are given by polynomials in the entries of the matrix. Since,

$$e(D) = \left\{ \begin{pmatrix} 0 & \cdots & 0 \\ \vdots & \ddots & \vdots \\ 0 & \cdots & 0 \end{pmatrix} \right\}$$

and since D is dense, by continuity this function must map the entire set of $n \times n$ matrices to the zero matrix. Therefore the Cayley–Hamilton theorem is true for complex numbers, and must therefore also hold for $\mathbb{Q}$ or $\mathbb{R}$ valued matrices.

While this provides a valid proof, the argument is not very satisfactory, since the identities represented by the theorem do not in any way depend on the nature of the matrix (diagonalizable or not), nor on the kind of entries allowed (for matrices with real entries the diagonalizable ones do not form a dense set, and it seems strange one would have to consider complex matrices to see that the Cayley–Hamilton theorem holds for them). We shall therefore now consider only arguments that prove the theorem directly for any matrix using algebraic manipulations only; these also have the benefit of working for matrices with entries in any commutative ring.

There is a great variety of such proofs of the Cayley–Hamilton theorem, of which several will be given here. They vary in the amount of abstract algebraic notions required to understand the proof. The simplest proofs use just those notions needed to formulate the theorem (matrices, polynomials with numeric entries, determinants), but involve technical computations that render somewhat mysterious the fact that they lead precisely to the correct conclusion. It is possible to avoid such details, but at the price of involving more subtle algebraic notions: polynomials with coefficients in a non-commutative ring, or matrices with unusual kinds of entries.

Adjugate Matrices

All proofs below use the notion of the adjugate matrix adj(*M*) of an *n*×*n* matrix *M*, the transpose of its cofactor matrix.

This is a matrix whose coefficients are given by polynomial expressions in the coefficients of *M* (in fact, by certain $(n-1)\times(n-1)$ determinants), in such a way that the following fundamental relations hold,

$$\mathrm{adj}(M)\cdot M = \det(M)I_n = M\cdot\mathrm{adj}(M).$$

These relations are a direct consequence of the basic properties of determinants: evaluation of the (*i*,*j*) entry of the matrix product on the left gives the expansion by column *j* of the determinant of the matrix obtained from *M* by replacing column *i* by a copy of column *j*, which is det(*M*) if *i* = *j* and zero otherwise; the matrix product on the right is similar, but for expansions by rows.

Being a consequence of just algebraic expression manipulation, these relations are valid for matrices with entries in any commutative ring (commutativity must be assumed for determinants to be defined in the first place). This is important to note here, because these relations will be applied below for matrices with non-numeric entries such as polynomials.

A Direct Algebraic Proof

This proof uses just the kind of objects needed to formulate the Cayley–Hamilton theorem: matrices with polynomials as entries. The matrix $t\,I_n - A$ whose determinant is the characteristic polynomial of *A* is such a matrix, and since polynomials form a commutative ring, it has an adjugate:

$$B = \mathrm{adj}(tI_n - A).$$

Then, according to the right-hand fundamental relation of the adjugate, one has:

$$(tI_n - A)B = \det(tI_n - A)I_n = p(t)I_n.$$

Since *B* is also a matrix with polynomials in *t* as entries, one can, for each *i*, collect the coefficients of t^i in each entry to form a matrix B_i of numbers, such that one has,

$$B = \sum_{i=0}^{n-1} t^i B_i.$$

(The way the entries of *B* are defined makes clear that no powers higher than t^{n-1} occur). While this *looks* like a polynomial with matrices as coefficients, we shall not consider such a notion; it is just a way to write a matrix with polynomial entries as a linear combination of *n* constant matrices, and the coefficient t^i has been written to the left of the matrix to stress this point of view.

Now, one can expand the matrix product in our equation by bilinearity,

$$p(t)I_n = (tI_n - A)B$$

$$= (tI_n - A)\sum_{i=0}^{n-1} t^i B_i$$

$$= \sum_{i=0}^{n-1} tI_n \cdot t^i B_i - \sum_{i=0}^{n-1} A \cdot t^i B_i$$

$$= \sum_{i=0}^{n-1} t^{i+1} B_i - \sum_{i=0}^{n-1} t^i AB_i$$

$$= t^n B_{n-1} + \sum_{i=1}^{n-1} t^i (B_{i-1} - AB_i) - AB_0 \ .$$

Writing,

$$p(t)I_n = t^n I_n + t^{n-1} c_{n-1} I_n + \cdots + t c_1 I_n + c_0 I_n \ ,$$

one obtains an equality of two matrices with polynomial entries, written as linear combinations of constant matrices with powers of t as coefficients.

Such an equality can hold only if in any matrix position the entry that is multiplied by a given power t^i is the same on both sides; it follows that the constant matrices with coefficient t^i in both expressions must be equal. Writing these equations then for i from n down to 0, one finds:

$$B_{n-1} = I_n, \qquad B_{i-1} - AB_i = c_i I_n \quad \text{for } 1 \le i \le n-1, \qquad -AB_0 = c_0 I_n \ .$$

Finally, multiply the equation of the coefficients of t^i from the left by A^i, and sum up:

$$A^n B_{n-1} + \sum_{i=1}^{n-1}\left(A^i B_{i-1} - A^{i+1} B_i\right) - AB_0 = A^n + c_{n-1} A^{n-1} + \cdots + c_1 A + c_0 I_n \ .$$

The left-hand sides form a telescoping sum and cancel completely; the right-hand sides add up to $p(A)$:

$$0 = p(A) \ .$$

This completes the proof.

A Proof using Polynomials with Matrix Coefficients

This proof is similar to the first one, but tries to give meaning to the notion of polynomial with matrix coefficients that was suggested by the expressions occurring in that proof. This requires considerable care, since it is somewhat unusual to consider polynomials with coefficients in a non-commutative ring, and not all reasoning that is valid for commutative polynomials can be applied in this setting.

Notably, while arithmetic of polynomials over a commutative ring models the arithmetic of polynomial functions, this is not the case over a non-commutative ring (in fact there is no obvious notion of polynomial function in this case that is closed under multiplication). So when considering polynomials in t with matrix coefficients, the variable t must not be thought of as an "unknown", but as a formal symbol that is to be manipulated according to given rules; in particular one cannot just set t to a specific value.

$$(f+g)(x) = \sum_i (f_i + g_i)x^i = \sum_i f_i x^i + \sum_i g_i x^i = f(x) + g(x).$$

Let $M(n, R)$ be the ring of $n \times n$ matrices with entries in some ring R (such as the real or complex numbers) that has A as an element. Matrices with as coefficients polynomials in t, such as $tI_n - A$ or its adjugate B in the first proof, are elements of $M(n, R[t])$.

By collecting like powers of t, such matrices can be written as "polynomials" in t with constant matrices as coefficients; write $M(n, R)[t]$ for the set of such polynomials. Since this set is in bijection with $M(n, R[t])$, one defines arithmetic operations on it correspondingly, in particular multiplication is given by:

$$\left(\sum_i M_i t^i\right)\left(\sum_j N_j t^j\right) = \sum_{i,j} (M_i N_j) t^{i+j},$$

respecting the order of the coefficient matrices from the two operands; obviously this gives a non-commutative multiplication.

Thus, the identity,

$$(tI_n - A)B = p(t)I_n.$$

from the first proof can be viewed as one involving a multiplication of elements in $M(n, R)[t]$.

At this point, it is tempting to simply set t equal to the matrix A, which makes the first factor on the left equal to the null matrix, and the right hand side equal to $p(A)$; however, this is not an allowed operation when coefficients do not commute. It is possible to define a "right-evaluation map" $ev_A : M[t] \to M$, which replaces each t^i by the matrix power A^i of A, where one stipulates that the power is always to be multiplied on the right to the corresponding coefficient.

But this map is not a ring homomorphism: the right-evaluation of a product differs in general from the product of the right-evaluations. This is so because multiplication of polynomials with matrix coefficients does not model multiplication of expressions containing unknowns: a product $Mt^i Nt^j = (M \cdot N)t^{i+j}$ is defined assuming that t commutes with N, but this may fail if t is replaced by the matrix A.

One can work around this difficulty in the particular situation at hand, since the above right-evaluation map does become a ring homomorphism if the matrix A is in the center of the ring of coefficients, so that it commutes with all the coefficients of the polynomials (the argument proving this is straightforward, exactly because commuting t with coefficients is now justified after evaluation).

Now, A is not always in the center of M, but we may replace M with a smaller ring provided it contains all the coefficients of the polynomials in question: I_n, A, and the coefficients B_i of the polynomial B. The obvious choice for such a subring is the centralizer Z of A, the subring of all matrices that commute with A; by definition A is in the center of Z.

This centralizer obviously contains I_n, and A, but one has to show that it contains the matrices B_i. To do this, one combines the two fundamental relations for adjugates, writing out the adjugate B as a polynomial:

$$\left(\sum_{i=0}^{m} B_i t^i\right)(tI_n - A) = (tI_n - A)\sum_{i=0}^{m} B_i t^i$$

$$\sum_{i=0}^{m} B_i t^{i+1} - \sum_{i=0}^{m} B_i A t^i = \sum_{i=0}^{m} B_i t^{i+1} - \sum_{i=0}^{m} A B_i t^i$$

$$\sum_{i=0}^{m} B_i A t^i = \sum_{i=0}^{m} A B_i t^i.$$

Equating the coefficients shows that for each i, we have $A B_i = B_i A$ as desired. Having found the proper setting in which ev_A is indeed a homomorphism of rings, one can complete the proof as suggested above:

$$\mathrm{ev}_A\left(p(t)I_n\right) = \mathrm{ev}_A((tI_n - A)B)$$

$$p(A) = \mathrm{ev}_A(tI_n - A)\cdot\mathrm{ev}_A(B)$$

$$p(A) = (AI_n - A)\cdot\mathrm{ev}_A(B) = O\cdot\mathrm{ev}_A(B) = O.$$

This completes the proof.

A Synthesis of the First Two Proofs

In the first proof, one was able to determine the coefficients B_i of B based on the right-hand fundamental relation for the adjugate only. In fact the first n equations derived can be interpreted as determining the quotient B of the Euclidean division of the polynomial $p(t)I_n$ on the left by the monic polynomial $I_n t - A$, while the final equation expresses the fact that the remainder is zero. This division is performed in the ring of polynomials with matrix coefficients. Indeed, even over a non-commutative ring, Euclidean division by a monic polynomial P is defined, and always produces a unique quotient and remainder with the same degree condition as in the commutative case, provided it is specified at which side one wishes P to be a factor.

To see that quotient and remainder are unique (which is the important part of the statement here), it suffices to write $PQ + r = PQ' + r'$ as $P(Q - Q') = r' - r$ and observe that since P is monic, $P(Q-Q')$ cannot have a degree less than that of P, unless $Q=Q'$.

But the dividend $p(t)I_n$ and divisor $I_n t - A$ used here both lie in the subring $(R[A])[t]$, where $R[A]$ is the subring of the matrix ring $M(n, R)$ generated by A: the R-linear span of all powers of A. Therefore, the Euclidean division can in fact be performed within that *commutative* polynomial ring, and of course it then gives the same quotient B and remainder 0 as in the larger ring; in particular this shows that B in fact lies in $(R[A])[t]$.

But, in this commutative setting, it is valid to set t to A in the equation,

$$p(t)I_n = (tI_n - A)B;$$

in other words, to apply the evaluation map,

$$\mathrm{ev}_A : (R[A])[t] \to R[A]$$

which is a ring homomorphism, giving,

$$p(A) = 0 \cdot \mathrm{ev}_A(B) = 0$$

just like in the second proof, as desired.

In addition to proving the theorem, the above argument tells us that the coefficients B_i of B are polynomials in A, while from the second proof we only knew that they lie in the centralizer Z of A; in general Z is a larger subring than $R[A]$, and not necessarily commutative. In particular the constant term $B_0 = \mathrm{adj}(-A)$ lies in $R[A]$. Since A is an arbitrary square matrix, this proves that $\mathrm{adj}(A)$ can always be expressed as a polynomial in A (with coefficients that depend on A).

In fact, the equations found in the first proof allow successively expressing $B_{n-1}, \ldots, B_1, B_0$ as polynomials in A, which leads to the identity,

$$\mathrm{adj}(-A) = \sum_{i=1}^{n} c_i A^{i-1},$$

valid for all $n \times n$ matrices, where,

$$p(t) = t^n + c_{n-1}t^{n-1} + \cdots + c_1 t + c_0$$

is the characteristic polynomial of A.

Note that this identity also implies the statement of the Cayley–Hamilton theorem: one may move $\mathrm{adj}(-A)$ to the right hand side, multiply the resulting equation (on the left or on the right) by A, and use the fact that:

$$-A \cdot \mathrm{adj}(-A) = \mathrm{adj}(-A) \cdot (-A) = \det(-A)I_n = c_0 I_n.$$

A Proof using Matrices of Endomorphisms

As was mentioned above, the matrix $p(A)$ in statement of the theorem is obtained by first evaluating the determinant and then substituting the matrix A for t; doing that substitution into the matrix $tI_n - A$ before evaluating the determinant is not meaningful. Nevertheless, it is possible to give an interpretation where $p(A)$ is obtained directly as the value of a certain determinant, but this requires a more complicated setting, one of matrices over a ring in which one can interpret both the entries $A_{i,j}$ of A, and all of A itself. One could take for this the ring $M(n, R)$ of $n \times n$ matrices over R, where the entry $A_{i,j}$ is realised as $A_{i,j}I_n$, and A as itself. But considering matrices with matrices as entries might cause confusion with block matrices, which is not intended, as that gives the wrong notion of determinant (recall that the determinant of a matrix is defined as a sum of products of its entries, and in the case of a block matrix this is generally not the same as the corresponding sum of products of its blocks!). It is clearer to distinguish A from the endomorphism φ of an n-dimensional vector space V (or free R-module if R is not a field) defined by it in a basis $e_1,\dots$, e_n, and to take matrices over the ring $\mathrm{End}(V)$ of all such endomorphisms. Then $\varphi \in \mathrm{End}(V)$ is a possible matrix entry, while A designates the element of $M(n, \mathrm{End}(V))$ whose i,j entry is endomorphism of scalar multiplication by $A_{i,j}$; similarly I_n will be interpreted as element of $M(n, \mathrm{End}(V))$. However, since $\mathrm{End}(V)$ is not a commutative ring, no determinant is defined on $M(n, \mathrm{End}(V))$; this can only be done for matrices over a commutative subring of $\mathrm{End}(V)$. Now the entries of the matrix $\varphi I_n - A$ all lie in the subring $R[\varphi]$ generated by the identity and φ, which is commutative. Then a determinant map $M(n, R[\varphi]) \to R[\varphi]$ is defined, and $\det(\varphi I_n - A)$ evaluates to the value $p(\varphi)$ of the characteristic polynomial of A at φ (this holds independently of the relation between A and φ); the Cayley–Hamilton theorem states that $p(\varphi)$ is the null endomorphism.

In this form, the following proof can be obtained from that of (which in fact is the more general statement related to the Nakayama lemma; one takes for the ideal in that proposition the whole ring R). The fact that A is the matrix of φ in the basis $e_1,\dots$, e_n means that:

$$\varphi(e_i) = \sum_{j=1}^{n} A_{j,i} e_j \quad \text{for } i = 1,\dots,n.$$

One can interpret these as n components of one equation in V^n, whose members can be written using the matrix-vector product $M(n, \mathrm{End}(V)) \times V^n \to V^n$ that is defined as usual, but with individual entries $\psi \in \mathrm{End}(V)$ and v in V being "multiplied" by forming $\psi(v)$; this gives:

$$\varphi I_n \cdot E = A^{\mathrm{tr}} \cdot E,$$

where $E \in V^n$ is the element whose component i is e_i (in other words it is the basis $e_1,\dots$, e_n of V written as a column of vectors). Writing this equation as:

$$(\varphi I_n - A^{\mathrm{tr}}) \cdot E = 0 \in V^n$$

one recognizes the transpose of the matrix $\varphi I_n - A$ considered above, and its determinant (as element of $M(n, R[\varphi])$) is also $p(\varphi)$. To derive from this equation that $p(\varphi) = 0 \in \mathrm{End}(V)$, one left-multiplies by the adjugate matrix of $\varphi I_n - A^{\mathrm{tr}}$, which is defined in the matrix ring $M(n, R[\varphi])$, giving:

$$
\begin{aligned}
0 &= \mathrm{adj}(\varphi I_n - A^{\mathrm{tr}}) \cdot ((\varphi I_n - A^{\mathrm{tr}}) \cdot E) \\
&= (\mathrm{adj}(\varphi I_n - A^{\mathrm{tr}}) \cdot (\varphi I_n - A^{\mathrm{tr}})) \cdot E \\
&= (\det(\varphi I_n - A^{\mathrm{tr}}) I_n) \cdot E \\
&= (p(\varphi) I_n) \cdot E;
\end{aligned}
$$

the associativity of matrix-matrix and matrix-vector multiplication used in the first step is a purely formal property of those operations, independent of the nature of the entries. Now component i of this equation says that $p(\varphi)(e_i) = 0 \in V$; thus $p(\varphi)$ vanishes on all e_i, and since these elements generate V it follows that $p(\varphi) = 0 \in \mathrm{End}(V)$, completing the proof.

One additional fact that follows from this proof is that the matrix A whose characteristic polynomial is taken need not be identical to the value φ substituted into that polynomial; it suffices that φ be an endomorphism of V satisfying the initial equations,

$$\varphi(e_i) = \sum_j A_{j,i} e_j$$

for *some* sequence of elements $e_1,...,e_n$ that generate V (which space might have smaller dimension than n, or in case the ring R is not a field it might not be a free module at all).

A bogus "proof": $p(A) = \det(AI_n - A) = \det(A - A) = 0$.

One persistent elementary but *incorrect* argument for the theorem is to "simply" take the definition:

$$p(\lambda) = \det(\lambda I_n - A)$$

and substitute A for λ, obtaining:

$$p(A) = \det(AI_n - A) = \det(A - A) = 0$$

There are many ways to see why this argument is wrong. First, in Cayley–Hamilton theorem, $p(A)$ is an $n \times n$ *matrix*. However, the right hand side of the above equation is the value of a determinant, which is a *scalar*. So they cannot be equated unless $n = 1$ (i.e. A is just a scalar). Second, in the expression $\det(\lambda I_n - A)$, the variable λ actually occurs at the diagonal entries of the matrix $\lambda I_n - A$. To illustrate, consider the characteristic polynomial in the previous example again:

$$\det\begin{pmatrix} \lambda - 1 & -2 \\ -3 & \lambda - 4 \end{pmatrix}.$$

If one substitutes the entire matrix A for λ in those positions, one obtains:

$$\det\begin{pmatrix} \begin{pmatrix} 1 & 2 \\ 3 & 4 \end{pmatrix} - 1 & -2 \\ -3 & \begin{pmatrix} 1 & 2 \\ 3 & 4 \end{pmatrix} - 4 \end{pmatrix},$$

in which the "matrix" expression is simply not a valid one. Note, however, that if scalar multiples of identity matrices instead of scalars are subtracted in the above, i.e. if the substitution is performed as:

$$\det\begin{pmatrix} \begin{pmatrix} 1 & 2 \\ 3 & 4 \end{pmatrix} - I_2 & -2I_2 \\ -3I_2 & \begin{pmatrix} 1 & 2 \\ 3 & 4 \end{pmatrix} - 4I_2 \end{pmatrix},$$

then the determinant is indeed zero, but the expanded matrix in question does not evaluate to $AI_n - A$; nor can its determinant (a scalar) be compared to $p(A)$ (a matrix). So the argument that $p(A) = \det(AI_n - A) = 0$ still does not apply.

Actually, if such an argument holds, it should also hold when other multilinear forms instead of determinant is used. For instance, if we consider the permanent function and define $q(\lambda) = \text{perm}(\lambda I_n - A)$, then by the same argument, we should be able to "prove" that $q(A) = 0$. But this statement is demonstrably wrong. In the 2-dimensional case, for instance, the permanent of a matrix is given by,

$$\text{perm}\begin{pmatrix} a & b \\ c & d \end{pmatrix} = ad + bc.$$

So, for the matrix A in the previous example,

$$q(\lambda) = \text{perm}(\lambda I_2 - A) = \text{perm}\begin{pmatrix} \lambda-1 & -2 \\ -3 & \lambda-4 \end{pmatrix}$$

$$= (\lambda-1)(\lambda-4) + (-2)(-3) = \lambda^2 - 5\lambda + 10.$$

Yet one can verify that,

$$q(A) = A^2 - 5A + 10I_2 = 12I_2 \neq 0.$$

One of the proofs for Cayley–Hamilton theorem above bears some similarity to the argument that $p(A) = \det(AI_n - A) = 0$. By introducing a matrix with non-numeric coefficients, one can actually let A live inside a matrix entry, but then AI_n is not equal to A, and the conclusion is reached differently.

Proofs using Methods of Abstract Algebra

Basic properties of Hasse–Schmidt derivations on the exterior algebra $A = \wedge M$ of some B-module M (supposed to be free and of finite rank) have been used by Gatto & Salehyan to prove the Cayley–Hamilton theorem.

Abstraction and Generalizations

The above proofs show that the Cayley–Hamilton theorem holds for matrices with entries in any commutative ring R, and that $p(\varphi) = 0$ will hold whenever φ is an endomorphism of an R module generated by elements $e_1,...,e_n$ that satisfies:

$$\varphi(e_j) = \sum a_{ij}e_i, \qquad j = 1,...,n.$$

This more general version of the theorem is the source of the celebrated Nakayama lemma in commutative algebra and algebraic geometry.

References

- Matrices: byjus.com, Retrieved 17 January, 2019

- Determinant-of-a-matrix, maths-determinants: toppr.com, Retrieved 08 July, 2019

- Grcar, Joseph F. (2011a), "How ordinary elimination became Gaussian elimination", Historia Mathematica, 38 (2): 163–218, arXiv:0907.2397, doi:10.1016/j.hm.2010.06.003

- Eigenstuff, calculus: math.hmc.edu, Retrieved 14 June, 2019

- Garrett, Paul B. (2007). Abstract Algebra. NY: Chapman and Hall/CRC. ISBN 978-1584886891

- Bernstein, Dennis (2005). Matrix Mathematics. Princeton University Press. p. 45. ISBN 978-0-691-11802-4

- Press, WH; Teukolsky, SA; Vetterling, WT; Flannery, BP (2007), "Section 2.3", Numerical Recipes: The Art of Scientific Computing (3rd ed.), New York: Cambridge University Press, ISBN 978-0-521-88068-8

Complex Analysis

The branch of mathematical analysis that examines functions of complex numbers is referred to as complex analysis. Some of the main concepts of complex analysis are Cauchy's integral theorem and residue theorem, complex function, analytic function, etc. This chapter has been carefully written to provide an easy understanding of various aspects of complex analysis.

Complex analysis is the study of complex numbers together with their derivatives, manipulation, and other properties. Complex analysis is an extremely powerful tool with an unexpectedly large number of practical applications to the solution of physical problems. Contour integration, for example, provides a method of computing difficult integrals by investigating the singularities of the function in regions of the complex plane near and between the limits of integration.

The key result in complex analysis is the Cauchy integral theorem, which is the reason that single-variable complex analysis has so many nice results. A single example of the unexpected power of complex analysis is Picard's great theorem, which states that an analytic function assumes every complex number, with possibly one exception, infinitely often in any neighborhood of an essential singularity.

A fundamental result of complex analysis is the Cauchy-Riemann equations, which give the conditions a function must satisfy in order for a complex generalization of the derivative, the so-called complex derivative, to exist. When the complex derivative is defined "everywhere," the function is said to be analytic.

COMPLEX FUNCTION

Let the complex variable z be defined by $z = x + iy$ where x and y are real variables and i is, as usual, given by $i^2 = -1$. Now let a second complex variable w be defined by $w = u + iv$ where u and v are real variables. If there is a relationship between w and z such that to each value of z in a given region of the z–plane there is assigned one, and only one, value of w then w is said to be a function of z, defined on the given region. In this case we write,

$$w = f(z).$$

As a example consider $w = z^2 - z$, which is defined for all values of z (that is, the right-hand side can be computed for every value of z). Then, remembering that $z = x + iy$,

$$w = u + iv = (x + iy)^2 - (x + iy) = x^2 + 2ixy - y^2 - x - iy.$$

Hence, equating real and imaginary parts: $u = x^2 - x - y^2$ and $v = 2xy - y$.

If $z = 2 + 3i$, for example, then $x = 2, y = 3$ so that $u = 4 - 2 - 9 = -7$ and $v = 12 - 3 = 9$, giving w = −7 + 9i.

Limit of a Function

The limit of $w = f(z)$ as $z \to z_0$ is a number ℓ such that $\ell |f(z) - \ell|$ can be made as small as we wish by making $|z - z_0|$ sufficiently small. In some cases the limit is simply $f(z_0)$,, as is the case for $w = z^2 - z$.. For example, the limit of this function as $z \to i$ is $f(i) = i^2 - i = -1 - i$.

There is a fundamental difference from functions of a real variable: there, we could approach a point on the curve $y = g(x)$ either from the left or from the right when considering limits of $g(x)$ at such points. With the function $f(z)$ we are allowed to approach the point $z = z_0$ along any path in the z-plane; we require merely that the distance $|z = z_0|$ decreases to zero.

Suppose that we want to find the limit of $f(z) = z^2 - z$ as $z \to 2 + i$ along each of the paths (a), (b) and (c) indicated in figure.

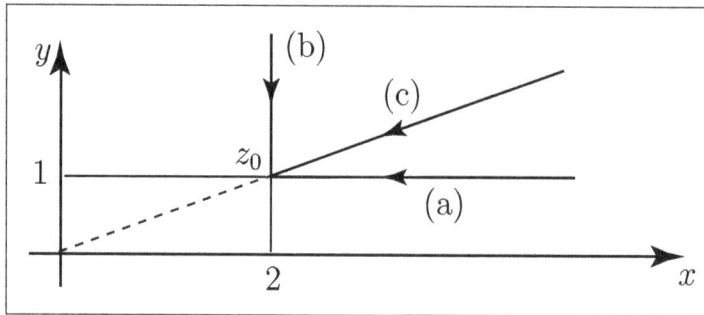

(a) Along this path $z = x + i$ (for any x) and $z^2 - z = x^2 + 2xi - 1 - x - i$ That is: $z^2 - z = x^2 - 1 - x + (2x - 1)i$.

As $z \to 2 + i$, then $x \to 2$ so that the limit of:

$z^2 - z$ is $2^2 - 1 - 2 + (4 - 1)i = 1 + 3i$.

(b) Here $z = 2 + yi$ (for any y) so that $z^2 - z = 4 - y^2 - 2 + (4y - y)i$. As $z \to 2 + i, y \to 1$ so that the limit of $z^2 - z$ is $4 - 1 - 2 + (4 - 1)i = 1 + 3i$.

(c) Here $z = k(2 + i)$ where k is a real number. Then:

$z^2 - z = k^2(4 + 4i - 1) - k(2 + i) = 3k^2 - 2k + (4k^2 - k)i$.

As $z \to 2 + i, k \to 1$ so that the limit of $z^2 - z$ is $3 - 2 + (4 - 1)i = 1 + 3i$.

In each case the limit is the same.

Consider the function $f(z) = \dfrac{\overline{z}}{z}$. Along the x-axis moving towards the origin from the right $z = x$ and $\overline{z} = x$ so that $f(z) = 1$ which is the limit as $z \to 0$ along this path.

However, we can approach the origin along any path. If instead we approach the origin along the positive y-axis z = iy then,

$$\overline{z} = -iy \text{ and } f(z) = \frac{\overline{z}}{z} = -1, \text{which is the limit as } z \to 0 \text{ along this path.}$$

Since these two limits are distinct then $\lim_{z \to 0} \frac{\overline{z}}{z}$ does not exist.

We cannot assume that the limit of a function $f(z)$ as $z \to z_0$ is independent of the path chosen.

The function $f(z)$ is continuous as $z \to z_0$ if the following two statements are true:

(a) $f(z_0)$ exists.

(b) $\lim_{z \to z_0} f(z)$ exists and is equal to $f(z_0)$.

As an example consider $f(z) = \frac{z^2 + 4}{z^2 + 9}$, As $z \to i$, then $f(z) \to f(i) = \frac{i^2 + 4}{i^2 + 9} = \frac{3}{8}$. Thus $f(z)$ is continuous at $z = i$.

However, when $z^2 + 9 = 0$ then $z = \pm 3i$ and neither $f(3i)$ nor $f(-3i)$ exists. Thus $\frac{z^2 + 4}{z^2 + 9}$ is discontinuous at $z = \pm 3i$. It is easily shown that these are the only points of discontinuity.

Differentiating Functions of a Complex Variable

The function f(z) is said to be differentiable at z = z_0 if:

$$\lim_{\Delta z \to 0} \left\{ \frac{f(z_0 + \Delta z) - f(z_0)}{\Delta z} \right\} \text{exists. Here } \Delta z = \Delta x + i\Delta y$$

Apart from a change of notation this is precisely the same as the definition of the derivative of a function of a real variable. Not surprisingly then, the rules of differentiation used in functions of a real variable can be used to differentiate functions of a complex variable. The value of the limit is the derivative of $f(z)$ at $z = z_0$ and is often denoted by $\frac{df}{dz}|z = z_0$ or by $f'(z_0)$.

A point at which the derivative does not exist is called a singular point of the function.

A function f(z) is said to be analytic at a point z_0 if it is differentiable throughout a neighbourhood of z_0, however small. (A neighbourhood of z_0 is the region contained within some circle $|x - z_0| = r$.)

For example, the function $f(z) = \frac{1}{z^2 + 1}$ has singular points where $z^2 + 1 = 0$, i.e. at $z = \pm i$.

For all other points the usual rules for differentiation apply and hence,

$$f'(z) = -\frac{2z}{(z^2 + 1)^2} \quad \text{(quotient rule)}$$

So, for example, at $z = 3i, f'(z) = -\frac{6i}{(-9 + 1)^2} = -\frac{3}{32}i.$

The simple function $f(z) = \bar{z} = x - iy$ is not analytic anywhere in the complex plane. To see this consider looking at the derivative at an arbitrary point z_0. We easily see that,

$$R = f(z_0 + \Delta z) - f(z_0)$$

$$= \frac{(x_0 + \Delta x) - i(y_0 + \Delta y) - (x_0 - iy_0)}{\Delta x + i\Delta y} = \frac{\Delta x - i\,\Delta y}{\Delta x + i\,\Delta y}$$

Hence f(z) will fail to have a derivative at z_0 if we can show that this expression has no limit. To do this we consider looking at the limit of the function along two distinct paths.

Along a path parallel to the x-axis:

$\Delta y = 0$ so that $R = \dfrac{\Delta x}{\Delta x} = 1$, and this is the limit as $\Delta z = \Delta x \rightarrow 0$.

Along a path parallel to the y-axis:

$\Delta x = 0$ so that $R = \dfrac{-i\,\Delta y}{i\Delta y} = -1$, and this is the limit as $\Delta z = \Delta y \rightarrow 0$.

As these two values of R are distinct, the limit of $\dfrac{f(z + \Delta z) - f(z)\Delta z}{\Delta z}$ as $z \rightarrow z_0$ does not exist and so f(z) fails to be differentiable at any point. Hence it is not analytic anywhere.

ANALYTIC FUNCTION

In mathematics, an analytic function is a function that is locally given by a convergent power series. There exist both real analytic functions and complex analytic functions, categories that are similar in some ways, but different in others. Functions of each type are infinitely differentiable, but complex analytic functions exhibit properties that do not hold generally for real analytic functions. A function is analytic if and only if its Taylor series about xo converges to the function in some neighborhood for every xo in its domain.

Formally, a function f is real analytic on an open set D in the real line if for any $x_0 \in D$ one can write

$$f(x) = \sum_{n=0}^{\infty} a_n (x - x_0)^n = a_0 + a_1(x - x_0) + a_2(x - x_0)^2 + a_3(x - x_0)^3 + \cdots$$

in which the coefficients $a_0, a_1, \ldots a_0, a_1, \ldots$ are real numbers and the series is convergent to $f(x)$ in a neighborhood of x_0.

Alternatively, an analytic function is an infinitely differentiable function such that the Taylor series at any point x_0 in its domain:

$$T(x) = \sum_{n=0}^{\infty} \frac{f^{(n)}(x_0)}{n!}(x - x_0)^n$$

converges to $f(x)$ in a neighborhood of x_0 pointwise. The set of all real analytic functions on a given set D is often denoted by $C^\omega(D)$.

A function f defined on some subset of the real line is said to be real analytic at a point x if there is a neighborhood D of x on which f is real analytic.

The definition of a complex analytic function is obtained by replacing, in the definitions above, "real" with "complex" and "real line" with "complex plane". A function is complex analytic if and only if it is holomorphic i.e. it is complex differentiable. For this reason the terms "holomorphic" and "analytic" are often used interchangeably for such functions.

Typical examples of analytic functions are:

- All elementary functions:

  ◦ All polynomials: if a polynomial has degree n, any terms of degree larger than n in its Taylor series expansion must immediately vanish to 0, and so this series will be trivially convergent. Furthermore, every polynomial is its own Maclaurin series.

  ◦ The exponential function is analytic. Any Taylor series for this function converges not only for x close enough to x_0 (as in the definition) but for all values of x (real or complex).

  ◦ The trigonometric functions, logarithm, and the power functions are analytic on any open set of their domain.

- Most special functions (at least in some range of the complex plane):

  ◦ Hypergeometric functions.

    ▪ Bessel functions.

  ◦ Gamma function.

Typical examples of functions that are not analytic are:

- The absolute value function when defined on the set of real numbers or complex numbers is not everywhere analytic because it is not differentiable at 0. Piecewise defined functions (functions given by different formulae in different regions) are typically not analytic where the pieces meet.

- The complex conjugate function z → z* is not complex analytic, although its restriction to the real line is the identity function and therefore real analytic, and it is real analytic as a function from $\mathbb{R}^2$ to $\mathbb{R}^2$.

- Other non-analytic smooth functions.

Alternative Characterizations

The following conditions are equivalent:

- f is real analytic on an open set D.

- There is a complex analytic extension of f to an open set $G \subset \mathbb{C}$ which contains D.

- f is real smooth and for every compact set $K \subset D$ there exists a constant C such that for every $x \in K$ and every non-negative integer k the following bound holds.

$$\left| \frac{d^k f}{dx^k}(x) \right| \leq C^{k+1} k!$$

Complex analytic functions are exactly equivalent to holomorphic functions, and are thus much more easily characterized.

For the case of an analytic function with several variables, the real analyticity can be characterized using the Fourier–Bros–Iagolnitzer transform. The third characterization has also a direct generalization for the multivariate case.

Properties of Analytic Functions

- The sums, products, and compositions of analytic functions are analytic.

- The reciprocal of an analytic function that is nowhere zero is analytic, as is the inverse of an invertible analytic function whose derivative is nowhere zero.

- Any analytic function is smooth, that is, infinitely differentiable. The converse is not true for real functions; in fact, in a certain sense, the real analytic functions are sparse compared to all real infinitely differentiable functions. For the complex numbers, the converse does hold, and in fact any function differentiable *once* on an open set is analytic on that set.

- For any open set $\Omega \subseteq C$, the set A(Ω) of all analytic functions u : $\Omega \rightarrow C$ is a Fréchet space with respect to the uniform convergence on compact sets. The fact that uniform limits on compact sets of analytic functions are analytic is an easy consequence of Morera's theorem. The set $A_\infty(\Omega)$ of all bounded analytic functions with the supremum norm is a Banach space.

A polynomial cannot be zero at too many points unless it is the zero polynomial (more precisely, the number of zeros is at most the degree of the polynomial). A similar but weaker statement holds for analytic functions. If the set of zeros of an analytic function f has an accumulation point inside its domain, then f is zero everywhere on the connected component containing the accumulation point. In other words, if (r_n) is a sequence of distinct numbers such that $f(r_n) = 0$ for all n and this sequence converges to a point r in the domain of D, then f is identically zero on the connected component of D containing r. This is known as the Principle of Permanence.

Also, if all the derivatives of an analytic function at a point are zero, the function is constant on the corresponding connected component.

These statements imply that while analytic functions do have more degrees of freedom than polynomials, they are still quite rigid.

Analyticity and Differentiability

As noted above, any analytic function (real or complex) is infinitely differentiable (also known as smooth, or C^∞). Note that this differentiability is in the sense of real variables; compare complex derivatives below. There exist smooth real functions that are not analytic. In fact there are many such functions.

The situation is quite different when one considers complex analytic functions and complex derivatives. It can be proved that any complex function differentiable (in the complex sense) in an open set is analytic. Consequently, in complex analysis, the term *analytic function* is synonymous with *holomorphic function*.

Real versus Complex Analytic Functions

Real and complex analytic functions have important differences (one could notice that even from their different relationship with differentiability). Analyticity of complex functions is a more restrictive property, as it has more restrictive necessary conditions and complex analytic functions have more structure than their real-line counterparts.

According to Liouville's theorem, any bounded complex analytic function defined on the whole complex plane is constant. The corresponding statement for real analytic functions, with the complex plane replaced by the real line, is clearly false; this is illustrated by:

$$f(x) = \frac{1}{x^2 + 1}.$$

Also, if a complex analytic function is defined in an open ball around a point x_0, its power series expansion at x_0 is convergent in the whole open ball (holomorphic functions are analytic). This statement for real analytic functions (with open ball meaning an open interval of the real line rather than an open disk of the complex plane) is not true in general; the function of the example above gives an example for $x_0 = 0$ and a ball of radius exceeding 1, since the power series $1 - x^2 + x^4 - x^6...$ diverges for $|x| > 1$.

Any real analytic function on some open set on the real line can be extended to a complex analytic function on some open set of the complex plane. However, not every real analytic function defined on the whole real line can be extended to a complex function defined on the whole complex plane. The function $f(x)$ defined in the paragraph above is a counterexample, as it is not defined for $x = \pm i$. This explains why the Taylor series of $f(x)$ diverges for $|x| > 1$, i.e., the radius of convergence is 1 because the complexified function has a pole at distance 1 from the evaluation point 0 and no further poles within the open disc of radius 1 around the evaluation point.

Analytic Functions of Several Variables

One can define analytic functions in several variables by means of power series in those variables. Analytic functions of several variables have some of the same properties as analytic functions of one variable. However, especially for complex analytic functions, new and interesting phenomena show up when working in 2 or more dimensions:

- Zero sets of complex analytic functions in more than one variable are never discrete due to Hartogs's extension theorem.

- Domains of holomorphy for single valued functions consist of arbitrary open sets. However, in several complex variables the characterization of domains of holomorphy leads to the notion of pseudoconvexity.

CAUCHY-RIEMANN EQUATIONS

Let,

$$f(x, y) \equiv u(x, y) + iv(x, y),$$

Where,

$$z \equiv x + iy,$$

So,

$$dz = dx + idy.$$

The total derivative of f with respect to z is then:

$$\frac{df}{dz} = \frac{\partial f}{\partial x}\frac{\partial x}{\partial z} + \frac{\partial f}{\partial y}\frac{\partial y}{\partial z}$$

$$= \frac{1}{2}\left(\frac{\partial f}{\partial x} - i\frac{\partial f}{\partial y}\right).$$

In terms of u and v, $\dfrac{1}{2}\left(\dfrac{\partial f}{\partial x} - i\dfrac{\partial f}{\partial y}\right)$ becomes:

$$\frac{df}{dz} = \frac{1}{2}\left[\left(\frac{\partial u}{\partial x} + i\frac{\partial v}{\partial v}\right) - i\left(\frac{\partial u}{\partial y} + i\frac{\partial v}{\partial y}\right)\right]$$

$$= \frac{1}{2}\left[\left(\frac{\partial u}{\partial x} + i\frac{\partial v}{\partial x}\right) + \left(-i\frac{\partial u}{\partial y} + \frac{\partial v}{\partial y}\right)\right].$$

Along the real, or x-axis, $\partial f / \partial y = 0$, so,

$$\frac{df}{dz} = \frac{1}{2}\left(\frac{\partial u}{\partial x} + i\frac{\partial v}{\partial x}\right).$$

Along the imaginary, or y-axis, $\partial f / \partial x = 0$, so,

$$\frac{df}{dz} = \frac{1}{2}\left(-i\frac{\partial u}{\partial y} + \frac{\partial v}{\partial y}\right).$$

If f is complex differentiable, then the value of the derivative must be the same for a given dz, regardless of its orientation. Therefore, $\frac{df}{dz} = \frac{1}{2}\left(\frac{\partial u}{\partial x} + i\frac{\partial v}{\partial x}\right)$ must equal $\frac{df}{dz} = \frac{1}{2}\left(-i\frac{\partial u}{\partial y} + \frac{\partial v}{\partial y}\right)$, which requires that:

$$\frac{\partial u}{\partial x} = \frac{\partial v}{\partial y}$$

and

$$\frac{\partial v}{\partial x} = \frac{\partial u}{\partial y}$$

These are known as the Cauchy-Riemann equations.

They lead to the conditions:

$$\frac{\partial^2 u}{\partial x^2} = \frac{\partial^2 u}{\partial y^2}$$

$$\frac{\partial^2 u}{\partial x^2} = \frac{\partial^2 v}{\partial y^2}$$

The Cauchy-Riemann equations may be concisely written as:

$$\frac{df}{dz} = \frac{1}{2}\left[\left(\frac{\partial f}{\partial x} + i\frac{\partial f}{\partial y^2}\right)\right]$$

$$= \frac{1}{2}\left[\left(\frac{\partial u}{\partial x} + i\frac{\partial v}{\partial x}\right) + i\left(\frac{\partial u}{\partial y} + i\frac{\partial v}{\partial y}\right)\right]$$

$$= \frac{1}{2}\left[\left(\frac{\partial u}{\partial x} - \frac{\partial v}{\partial y}\right) + i\left(\frac{\partial u}{\partial y} + \frac{\partial v}{\partial x}\right)\right] = 0.$$

where z is the complex conjugate.

If $z = r\, e^{i\theta}$, then the Cauchy-Riemann equations become:

$$\frac{\partial u}{\partial r} = \frac{1}{r}\frac{\partial v}{\partial \theta}$$

$$\frac{1}{r}\frac{\partial u}{\partial \theta} = -\frac{\partial v}{\partial r}$$

If u and v satisfy the Cauchy-Riemann equations, they also satisfy Laplace's equation in two dimensions, since:

$$\frac{\partial^2 u}{\partial x^2} + \frac{\partial^2 u}{\partial y^2} = \frac{\partial}{\partial x}\left(\frac{\partial v}{\partial y}\right) + \frac{\partial}{\partial y}\left(-\frac{\partial v}{\partial x}\right) = 0$$

$$\frac{\partial^2 v}{\partial x^2} + \frac{\partial^2 v}{\partial y^2} = \frac{\partial}{\partial x}\left(-\frac{\partial u}{\partial y}\right) + \frac{\partial}{\partial y}\left(-\frac{\partial u}{\partial x}\right) = 0.$$

By picking an arbitrary f(z), solutions can be found which automatically satisfy the Cauchy-Riemann equations and Laplace's equation. This fact is used to use conformal mappings to find solutions to physical problems involving scalar potentials such as fluid flow and electrostatics.

CAUCHY'S INTEGRAL THEOREM

The Cauchy integral theorem is a concept of complex analysis in mathematics. It is a very important statement for holomorphic functions about line integrals in the complex plane.

Cauchy's integral theorem states that, "if there is an open set say U, of C, the complex number set, which is easily connected and let there be defined a holomorphic function f such that f : U → C. Also, let ? in U be a rectifiable path where the start and end point are same. In that case, we have,

$$\oint_\gamma f(z)\,dz = 0$$

Here, z = x + i y.

If we take into consideration the continuity of the partial derivatives of the holomorphic function, then we can prove the Cauchy integral theorem directly as a consequence of Green's theorem along with the fact that the imaginary and real parts of the function 'f' will satisfy the equations given by Cauchy-Riemann in the region that is bounded by γ with the open neighborhood U of γ region.

Let us break the integrand function 'f' and the differential dz into their respective real and imaginary parts.

$$f = a + i\,b$$

$$dz = dx + idy$$

So we get,

$$\oint_\gamma \left[f(z)\,dz \right] = \oint_\gamma (a + i\,b)(dx + i\,dy)$$

$$\Rightarrow \oint_\gamma \left[f(z)\,dz \right] = \oint_\gamma (a\,dx - b\,dy) + i \oint_\gamma (b\,dx + a\,dy)$$

From Green's theorem, replace the integrals around ? that is the closed contour by the area integral that is throughout the domain D which is again enclosed by γ. Then we get,

$$\oint_\gamma (a\,dx - b\,dy) = \iint_D \left(-\frac{\partial b}{\partial x} - \frac{\partial a}{\partial y} \right) dx\,dy$$

$$\oint_\gamma (b\,dx - a\,dy) = \iint_D \left(-\frac{\partial a}{\partial x} - \frac{\partial b}{\partial y} \right) dx\,dy$$

Since, they are the real and imaginary parts of a holomorphic function in domain D, therefore, a and b will satisfy the Cauchy-Riemann equations that are,

$$\frac{\partial a}{\partial x} = \frac{\partial b}{\partial y}$$

$$\frac{\partial b}{\partial x} = -\frac{\partial a}{\partial y}$$

From this we get,

$$\iint_D \left(-\frac{\partial b}{\partial x} - \frac{\partial a}{\partial y} \right) dx\,dy = \iint_D \left(-\frac{\partial b}{\partial x} + \frac{\partial b}{\partial x} \right) dx\,dy = 0$$

$$\iint_D \left(-\frac{\partial a}{\partial x} - \frac{\partial b}{\partial y} \right) dx\,dy = \iint_D \left(-\frac{\partial a}{\partial x} + \frac{\partial a}{\partial x} \right) dx\,dy = 0$$

From all the above we get that,

$$\oint_\gamma \left[f(z)\,dz \right] = 0$$

Formula Back to Top

If we have a circle C in a complex plane that has a positive orientation, 'f' is a holomorphic function on an open set that contains circle C along with its interior and 'z' is any point that lies in interior of circle C then we have,

$$f(z) = \frac{1}{(2\pi i)} \int_C \left[f\frac{u}{u-z} \right] du$$

Derivative of an Analytic Function

A function that can be given by a convergent power series locally is known as an analytic function. We have both complex and real type of analytic functions. The function above is a type of analytic function only, that is $f(z) = \frac{1}{(2\pi i)} \int_C \left[f\frac{u}{u-z} \right] du$ is analytic when we have,

$$f'(z) = \frac{1}{(2\pi i)} \int_C \left[f\frac{u}{(u-z)^2} \right] du$$

for any 'z' lying in the interior of C which is a simple closed positively oriented contour. Also, R be a region that contains C, then, in this case, we say that function 'f' is analytic in R.

Let us see some examples on Cauchy's integral theorem for better understanding.

Example: If k is a real value, then prove that $\left| e^{(2k\pi i)} - i \right| \le 2\pi \left| k \right|$.

Let $f(b) = e^{(ikb)}$ where 'k' and 'b' are real.

This gives us:

$$\left| \int_0^{2\pi} e^{(ikb)} db \right| \le \int_0^{\left(2\pi\right)} \left| e^{(ikb)} \right| db = 2\pi$$

Also,

$$\left| \int_0^{2\pi} e^{(ikb)} db \right| = \left| \frac{e(ikb)}{ik} \right|_0^{2\pi} = \frac{\left| e^{(2k\pi i)} - 1 \right|}{\left| k \right|}$$

Combining both we get,

$$\frac{\left| e(2k\pi i) \right| - 1}{\left| k \right|} \le 2\pi$$

$$\Rightarrow \left| e^{2k\pi i} - 1 \right| \le 2\pi \left| k \right|$$

Hence, it is proved.

Example: Given a circle C with positive orientation |j| = 1. In this case we have,

$$\int_C \left[\frac{e^{(3j)}}{j^4} \right] dj = \frac{3^3 e^0 2\pi i}{3!} = 9\pi i$$

Example: If C is a circle with |t − i| = 1, then we have:

$$\int_C \left[\frac{1}{\left(t^2 + 1\right)^2} \right] dt$$

$$\int_C \left[\left(\frac{\frac{1}{\left(t+i\right)^2}}{\left(t-i\right)^2} \right) \right] dt$$

$$= 2\pi i \left(-2\left(i+i\right)^{(-3)} \right)$$

$$= -4 \pi i \left(\frac{1}{8 i^3} \right)$$

$$= \frac{\pi}{2}$$

RESIDUE

In mathematics, more specifically complex analysis, the residue is a complex number proportional to the contour integral of a meromorphic function along a path enclosing one of its singularities. (More generally, residues can be calculated for any function $f : \mathbb{C} \setminus \{a_k\}_k \to \mathbb{C}$ that is holomorphic except at the discrete points k, even if some of them are essential singularities). Residues can be computed quite easily and, once known, allow the determination of general contour integrals via the residue theorem.

The residue of a meromorphic function f at an isolated singularity $\mathrm{Res}(f, a)$ or $\mathrm{Res}_a(f)$, , is the unique value R such that $f(z) - R/(z-a)$ has an analytic antiderivative in a punctured disk $0 < |z - a| < \delta$.

Alternatively, residues can be calculated by finding Laurent series expansions, and one can define the residue as the coefficient a_{-1} of a Laurent series.

The definition of a residue can be generalized to arbitrary Riemann surfaces. Suppose ω is a 1-form on a Riemann surface. Let ω be meromorphic at some point x, so that we may write ω in local coordinates as $f(z)\, dz$. Then the residue of ω at x is defined to be the residue of $f(z)$ at the point corresponding to x.

Residue of a Monomial

Computing the residue of a monomial:

$$\oint_C z^k dz$$

makes most residue computations easy to do. Since path integral computations are homotopy invariant, we will let C be the circle with radius 1. Then, using the change of coordinates $z \to e^{i\theta}$ we find that:

$$dz \to d(e^{i\theta}) = i e^{i\theta}\, d\theta$$

hence our integral now reads as:

$$\oint_C z^k dz = \int_0^{2\pi} i e^{i(k+1)\theta}\, d\theta = \begin{cases} 2\pi i & \text{if } k = -1, \\ 0 & \text{otherwise.} \end{cases}$$

Application of Monomial Residue

As an example, consider the contour integral:

$$\oint_C \frac{e^z}{z^5} \, dz$$

where C is some simple closed curve about 0.

Let us evaluate this integral using a standard convergence result about integration by series. We can substitute the Taylor series for e^z into the integrand. The integral then becomes:

$$\oint_C \frac{1}{z^5} \left(1 + z + \frac{z^2}{2!} + \frac{z^3}{3!} + \frac{z^4}{4!} + \frac{z^5}{5!} + \frac{z^6}{6!} + \cdots \right) dz.$$

Let us bring the $1/z^5$ factor into the series. The contour integral of the series then writes:

$$\oint_C \left(\frac{1}{z^5} + \frac{z}{z^5} + \frac{z^2}{2!z^5} + \frac{z^3}{3!z^5} + \frac{z^4}{4!z^5} + \frac{z^5}{5!z^5} + \frac{z^6}{6!z^5} + \cdots \right) dz$$

$$\oint_C \left(\frac{1}{z^5} + \frac{1}{z^4} + \frac{1}{2!z^3} + \frac{1}{3!z^2} + \frac{1}{4!z} + \frac{1}{5!} + \frac{z}{6!} + \cdots \right) dz.$$

Since the series converges uniformly on the support of the integration path, we are allowed to exchange integration and summation. The series of the path integrals then collapses to a much simpler form because of the previous computation. So now the integral around C of every other term not in the form cz^{-1} is zero, and the integral is reduced to:

$$\oint_C \frac{1}{4!z} \, dz = \frac{1}{4!} \oint_C \frac{1}{z} \, dz = \frac{1}{4!} (2\pi i) = \frac{\pi i}{12}.$$

The value $1/4!$ is the *residue* of e^z/z^5 at $z = 0$, and is denoted:

$$\text{Res}_0 \frac{e^z}{z^5}, \text{ or } \text{Res}_{z=0} \frac{e^z}{z^5}, \text{ or } \text{Res}(f,0) \text{ for } f = \frac{e^z}{z^5}.$$

Calculating Residues

Suppose a punctured disk $D = \{z : 0 < |z - c| < R\}$ in the complex plane is given and f is a holomorphic function defined (at least) on D. The residue $\text{Res}(f, c)$ of f at c is the coefficient a_{-1} of $(z - c)^{-1}$ in the Laurent seriesexpansion of f around c. Various methods exist for calculating this value, and the choice of which method to use depends on the function in question, and on the nature of the singularity.

According to the residue theorem, we have:

$$\text{Res}(f,c) = \frac{1}{2\pi i} \oint_\gamma f(z) \, dz$$

where γ traces out a circle around c in a counterclockwise manner. We may choose the path γ to be a circle of radius ε around c, where ε is as small as we desire. This may be used for calculation in cases where the integral can be calculated directly, but it is usually the case that residues are used to simplify calculation of integrals, and not the other way around.

Removable Singularities

If the function f can be continued to a holomorphic function on the whole disk $|y - c| < R$, then Res(f, c) = 0. The converse is not generally true.

Simple Poles

At a simple pole c, the residue of f is given by:

$$\text{Res}(f,c) = \lim_{z \to c}(z-c)f(z).$$

It may be that the function f can be expressed as a quotient of two functions, f(z) = g(z)/h(z), where g and h are holomorphic functions in a neighbourhood of c, with h(c) = 0 and h'(c) ≠ 0. In such a case, L'Hôpital's rule can be used to simplify the above formula to:

$$\text{Res}(f,c) = \lim_{z \to c}(z-c)f(z) = \lim_{z \to c}\frac{zg(z)-cg(z)}{h(z)}$$

$$= \lim_{z \to c}\frac{g(z)+zg'(z)-cg'(z)}{h'(z)} = \frac{g(c)}{h'(c)}.$$

Limit Formula for Higher Order Poles

More generally, if c is a pole of order n, then the residue of f around z = c can be found by the formula:

$$\text{Res}(f,c) = \frac{1}{(n-1)!}\lim_{z \to c}\frac{d^{n-1}}{dz^{n-1}}\left((z-c)^n f(z)\right).$$

This formula can be very useful in determining the residues for low-order poles. For higher order poles, the calculations can become unmanageable, and series expansion is usually easier. For essential singularities, no such simple formula exists, and residues must usually be taken directly from series expansions.

Residue at Infinity

In general, the residue at infinity is given by:

$$\text{Res}(f(z),\infty) = -\text{Res}\left(\frac{1}{z^2}f\left(\frac{1}{z}\right),0\right).$$

If the following condition is met:

$$\lim_{|z|\to\infty} f(z) = 0,$$

then the residue at infinity can be computed using the following formula:

$$\text{Res}(f,\infty) = -\lim_{|z|\to\infty} z \cdot f(z).$$

If instead,

$$\lim_{|z|\to\infty} f(z) = c \ne 0,$$

then the residue at infinity is,

$$\text{Res}(f,\infty) = \lim_{|z|\to\infty} z^2 \cdot f'(z).$$

Series Methods

If parts or all of a function can be expanded into a Taylor series or Laurent series, which may be possible if the parts or the whole of the function has a standard series expansion, then calculating the residue is significantly simpler than by other methods.

1. As a first example, consider calculating the residues at the singularities of the function,

$$f(z) = \frac{\sin z}{z^2 - z}$$

which may be used to calculate certain contour integrals. This function appears to have a singularity at z = 0, but if one factorizes the denominator and thus writes the function as:

$$f(z) = \frac{\sin z}{z(z-1)}$$

it is apparent that the singularity at z = 0 is a removable singularity and then the residue at z = 0 is therefore 0.

The only other singularity is at z = 1. Recall the expression for the Taylor series for a function g(z) about z = a:

$$g(z) = g(a) + g'(a)(z-a) + \frac{g''(a)(z-a)^2}{2!} + \frac{g'''(a)(z-a)^3}{3!} + \cdots$$

So, for g(z) = sin z and a = 1 we have,

$$\sin z = \sin 1 + \cos 1(z-1) + \frac{-\sin 1(z-1)^2}{2!} + \frac{-\cos 1(z-1)^3}{3!} + \cdots.$$

and for g(z) = 1/z and a = 1 we have,

$$\frac{1}{z} = \frac{1}{(z-1)+1} = 1-(z-1)+(z-1)^2-(z-1)^3+\cdots.$$

Multiplying those two series and introducing 1/(z – 1) gives us,

$$\frac{\sin z}{z(z-1)} = \frac{\sin 1}{z-1} + (\cos 1 - \sin 1) + (z-1)\left(-\frac{\sin 1}{2!} - \cos 1 + \sin 1\right) + \cdots.$$

So the residue of f(z) at z = 1 is sin 1.

2. The next example shows that, computing a residue by series expansion, a major role is played by the Lagrange inversion theorem. Let,

$$u(z) := \sum_{k\geq 1} u_k z^k$$

be an entire function, and let,

$$v(z) := \sum_{k\geq 1} v_k z^k$$

with positive radius of convergence, and with $v_1 \neq 0$. So $v(z)$ has a local inverse $V(z)$ at 0, and $u(1/V(z))$ is meromorphic at 0. Then we have:

$$\text{Res}_0\big(u(1/V(z))\big) = \sum_{k=0}^{\infty} k u_k v_k.$$

Indeed,

$$\text{Res}_0\big(u(1/V(z))\big) = \text{Res}_0\left(\sum_{k\geq 1} u_k V(z)^{-k}\right) = \sum_{k\geq 1} u_k \, \text{Res}_0\big(V(z)^{-k}\big)$$

because the first series converges uniformly on any small circle around 0. Using the Lagrange inversion theorem:

$$\text{Res}_0\big(V(z)^{-k}\big) = k v_k,$$

and we get the above expression. For example, if $u(z) = z + z^2$ and also $v(z) = z + z^2$ then,

$$V(z) = \frac{2z}{1+\sqrt{1+4z}}$$

and

$$u(1/V(z)) = \frac{1+\sqrt{1+4z}}{2z} + \frac{1+2z+\sqrt{1+4z}}{2z^2}.$$

The first term contributes 1 to the residue, and the second term contributes 2 since it is asymptotic to $1/z^2 + 2/z$.

Note that, with the corresponding stronger symmetric assumptions on u(z) and v(z), it also follows:

$$\mathrm{Res}_0\left(u(1/V)\right) = \mathrm{Res}_0(v(1/U)),$$

where U(z) is a local inverse of u(z) at 0.

CAUCHY'S RESIDUE THEOREM

Cauchy's residue theorem is a consequence of Cauchy's integral formula:

$$f(z_0) = \frac{1}{2\pi i}\oint_C \frac{f(z)}{z - z_0}dz,$$

where f is an analytic function and C is a simple closed contour in the complex plane enclosing the point z_0 with positive orientation which means that it is traversed counter clock wise.

Since the integrand is analytic except for z = z_0, the integral is equal to the same integral with C replaced by a small circle inside the contour C with center z_0. This implies that we have with $z = z_0 + r\,e^{i\theta}$ and $0 \le \theta \le 2\pi$:

$$\oint_C \frac{f(z)}{z-z_0}dz = \lim_{r\downarrow 0}\int_0^{2\pi}\frac{f\left(z_0+re^{i\theta}\right)}{re^{i\theta}}ire^{i\theta}d\theta$$

$$= i\,f(z_0)\int_0^{2\pi}d\theta = 2\pi\,i\,f(z_0),$$

which proves Cauchy's integral formula.

This formula can be iterated to:

$$f^{(n)}(z_0) = \frac{n!}{2\pi i}\oint_C \frac{f(z)}{(z-z_0)}dz,\ \ n\in\mathbb{N}.$$

If f is an analytic function except for an isolated singularity at z = z_0, then f has a Laurent series representation of the form:

$$f(z) = \sum_{n=-\infty}^{\infty} a_n(z-z_0)^n.$$

Then the coefficient a_{-1} is called the residue of f at z = z_0. We use the notation:

$$\mathrm{Res}_f(z_0) = a_{-1}.$$

Now we have Cauchy's residue theorem.

Theorem: If C is a simple closed, positively oriented contour in the complex plane and f is analytic except for some points $z_1, z_2,..., z_n$ inside the contour C, then:

$$\oint_C f(z)\,dz = 2\pi i \sum_{k=1}^{n} \operatorname{Res}_f (z_k).$$

If f has a removable singularity at z = z_0, then the residue is equal to zero. If f has a single pole at z = z_0, then:

$$\operatorname{Res}_f (z_0) = \lim_{z \to z_0} (z - z_0) f(z)$$

and if f has a pole of order k at z = z_0, then:

$$\operatorname{Res}_f (z_0) = \frac{1}{(k-1)!} \lim_{z \to z_0} \frac{d^k}{dz^k} \left\{ (z - z_0)^k f(z) \right\}, \ \ k \in \{1, 2, 3, ...\}.$$

Cauchy's residue theorem can be used to compute real integrals by applying an appropriate contour in the complex plane.

References

- ComplexAnalysis: mathworld.wolfram.com, Retrieved 14 February, 2019

- Cauchys-integral-theorem, calculus: math.tutorvista.com, Retrieved 09 August, 2019

- Marsden, Jerrold E.; Hoffman, Michael J. (1998). Basic Complex Analysis (3rd ed.). W. H. Freeman. ISBN 978-0-7167-2877-1

- Cauchy-RiemannEquations: mathworld.wolfram.com, Retrieved 17 June, 2019

- Krantz, Steven; Parks, Harold R. (2002). A Primer of Real Analytic Functions (2nd ed.). Birkhäuser. ISBN 0-8176-4264-1

INDEX

P
Partial Derivative, 22-24, 31, 69
Partial Differential Equation, 89, 92, 110-111, 113-114, 117
Polynomial Expression, 164
Polynomial Functions, 38, 173
Power Series Method, 102, 105
Probability Density Function, 36

Q
Quadratic Equation, 8-10

R
Residue Theorem, 77, 181, 193-194, 198-199
Riemann Integrable, 58-60
Riemann Sum, 74-75, 77-78
Rolle's Theorem, 27-28, 30-31
Roots Of Unity, 14-16

S
Scalar Potential, 64
Scalars, 1-3, 79, 151, 178
Square Matrix, 131, 134, 137-139, 147, 151, 153-155, 157, 159, 162-163, 167, 176
Standard Convergence, 194
Stationary Points, 37-38
Surface Integral, 61, 79-81, 83, 86-87

T
Taylor Formula, 22, 26
Taylor Series, 41-44, 46-47, 50-53, 62, 92, 104, 184-185, 187, 194, 196
Trigonometric Functions, 44, 49, 185
Triple Scalar Product, 5-6

V
Vector Algebra, 1-2
Vector Calculus, 63, 73, 78, 83, 86
Vector Product, 4, 177

www.ingramcontent.com/pod-product-compliance
Lightning Source LLC
Chambersburg PA
CBHW082021190326
41458CB00010B/3233